ENVIRONMENTAL CONCERNS

An Inter-disciplinary Exercise

Proceedings of an international symposium held in Elsinore, Denmark, 17-20 September 1990.

Programme and Editorial Committee:

Niels E. Busch
Danish Technological Institute

Jon Fjeldså
University of Copenhagen

Jens Aage Hansen, Chairman
University of Aalborg, Denmark

Ebba Lund
*Royal Veterinary and Agricultural
University of Copenhagen*

Olaf Olsen
National Museum of Denmark

Erik Ib Schmidt
Former Under Secretary of State, Denmark

Assistant Editor:

Susanne Bro
*Danish Academy of Technical Sciences,
266 Lundtoftevej, Lyngby, Denmark*

Acknowledgements:

The symposium and the book publication were sponsored financially by The Rockwool Foundation, Denmark. The Danish Academy of Technical Sciences organized the symposium and undertook the management of the editorial work on the book.

ENVIRONMENTAL CONCERNS

An Inter-disciplinary Exercise

Edited by

J.Aa. Hansen

Department of Civil Engineering,
University of Aalborg,
Aalborg, Denmark

Sponsored by:
The Rockwool Foundation,
Denmark

Organized by:
Danish Academy of Technical Sciences,
Lyngby, Denmark

SPRINGER SCIENCE+BUSINESS MEDIA, B.V.

WITH 21 TABLES AND 41 ILLUSTRATIONS

© 1991 SPRINGER SCIENCE+BUSINESS MEDIA DORDRECHT
ORIGINALLY PUBLISHED BY ELSEVIER SCIENCE PUBLISHER LTD IN 1991
SOFTCOVER REPRINT OF THE HARDCOVER 1ST EDITION 1991

© 1991 MICHAEL THOMPSON—pp.243-256

British Library Cataloguing in Publication Data

Environmental concerns : an inter-disciplinary
exercise.
I. Hanson, Jens Aa.
333.7

ISBN 978-94-010-5291-7   ISBN 978-94-011-2904-6 (eBook)
DOI 10.1007/978-94-011-2904-6

Library of Congress CIP data applied for

PREFACE

"How am I concerned about the environment from where I stand professionally?" Or, "what is the environmental edge of my profession?" A number of outstanding academics were asked to consider these questions in 1989 with the aim of opening an innovative discussion on environmental changes and an understanding of how these are related to human activities.

At the personal invitation of the Programme Committee, 17 individuals, each of whom have different professional backgrounds and nationalities, prepared a paper and met for a symposium in Elsinore, Denmark, in September of 1990. Some of the authors revised their original papers as a result of the discussions during the symposium. After peer review and editing, the contributions were printed in this book in versions that express the authors' personal convictions and priority environmental concerns.

The authors contributed to the symposium by delivering papers and participating in stimulating and sometimes provocative discussions. The Programme Committee is grateful for the professional inspiration provided and for the spiritual and warm, social atmosphere in which the symposium took place.

The Programme Committee has prepared the introductory chapter. Much of the inspiration stems from the written contributions of the invited authors and from the discussions during the symposium. However, instead of trying to summarize the symposium contributions, the Programme Committee has attempted to focus on some of the major environmental problems and issues which were discussed during the symposium and discover why it has been so difficult to implement purposeful management. In addition, suggestions to improve the understanding of environmental changes are made; special interest lies in those changes which can be controlled or partially avoided by intelligent management. In this way the Programme Committee wants to supplement as well as to motivate the reading of the contributions of the invited authors.

The Rockwool Foundation has actively supported the symposium concept and has financially sponsored both the symposium and the publication of this book. The Danish Academy of Technical Sciences organized the symposium efficiently and in a way that greatly supported the professional, cultural and social activities. Both institutions deserve our appreciation and gratitude.

The Programme Committee
Copenhagen, April, 1991.

CONTENTS

Page

1.

ENVIRONMENTAL CHANGE - PROBLEMS AND OPTIONS

N.E. Busch, J. Fjeldså, J.Aa. Hansen, E. Lund, O. Olsen, E.I. Schmidt

INTRODUCTION

Environmental change is a global process which would also take place in the absence of man. What causes concern today is the pace of the change and the relationship between the presence of man and his deleterious impacts on the environment: e.g. change of climate, pollution of water, soil and air, over-exploitation of forests, and the extinction of animal and plant species. The concerns over these environmental changes are now shared by scientists, politicians and citizens and have set the agenda for meetings at all levels, nationally and internationally, to an extent which has never occurred before.

There are several problems involved in trying to turn such concerns into positive action. One problem is that the existing scientific knowledge will normally only allow the prediction of environmental impact with a degree of uncertainty. Another problem is that political decisions are based on opinion, which implies that existing scientific knowledge is sometimes disregarded or used to postpone political decisions if the scientific knowledge conflicts with topical, political issues.

Consequently, better environmental decision-making must rely on a better understanding and knowledge of the natural and technical sciences as well as the social and humanistic sciences. Combinations of these scientific resources may offer the tools for wiser management of environmental changes in the future.

ENVIRONMENT, HEALTH AND GLOBAL POPULATION

In order to evaluate the environment, health and the global population in context, it may be useful to distinguish between the developing and the industrialized countries.

Developing Countries

Most developing countries suffer severe health problems. Water-borne infectious diseases are frequent and the problem is on the increase because the population growth rate exceeds that of extending water supply and sanitation systems. Improved sanitation and water supply are prerequisites for alleviating the present problems and starting a better development but such action requires technology, financing, management and the mobilization of human resources with respect to cultural heritage, religion, education and community participation: i.e., no simple technology-fix can tackle this problem.

Another basic problem facing developing countries is the lack of energy and food. Again, the growth of the population and the concentration in big cities demand increased and steady supplies, which are often not available locally or nationally. There are several reasons for this: e.g. disproportionate population growth compared to the supply capabilities of land and water resources; inadequate farming and cropping techniques; or the lack of facilities for safe storage and preservation of food stuff.

A third problem in developing countries is the lack of training in basic hygiene and health care; poor sanitation and environmental management to prevent the spreading of infectious diseases; or the insufficient, or even inappropriate, supply of medicines to cure diseases.

Poor housing is a fourth problem related to the environment as well as to the health in developing countries. Slums are prevalent in many big cities, e.g. as large squatting areas with shacks and leaky shelters which are often situated in swamp or other areas not suitable for housing.

In the developing world, population growth is significant and it will increase further, at least initially, because of decreased infant mortality, improved sanitation, water supply, food supply, medical care and housing. This, in turn, will have an impact on the environment, both regionally and globally, and, thereby, on the life quality for each citizen on Earth.

Industrialized Countries

On the same topics the industrialized countries have different stories.

Drinking water may be of variable quality and give rise to concern as to its odour, taste, chemical and even microbial composition. These qualities, however, in rural as well as urban areas, are monitored and technology is available to remedy problems before they turn into fatalities.

Food supply is of ample quantity in the industrialized countries but quality is a matter of concern. A major food quality problem is residues which result from use of chemicals (biocides) in agriculture. These biocides are used to increase yields. However, as a result of modern biotechnology, new biochemicals may appear and, together with biological pest control, the existing use of chemical biocides may be substituted.

Work is being done in the industrialized countries in order to develop new and improve pharmaceutical products to prevent or cure diseases. Dietary research is also involved as part of prophylactic health care schemes. Compared to the developing countries, infectious diseases seem well under control.

For various reasons, it has not been possible to include AIDS in the discussion of this book, although the problem links developing and industrialized countries on a common issue of relevance to both health and the environment on a global scale.

In the industrialized countries, slums are also found in big cities, although in a different sense. Poverty, also in well-to-do industrialized countries, is a real and seemingly growing problem at local, regional and national levels. Obviously the unequal distribution of wealth is a problem for which the political and economical structures are still deficient, even in the most affluent countries. In addition, poorly designed and monotonous building structures are aesthetically unpleasant and dominating while badly planned roads facilitate traffic, but create noise and air pollution. Sometimes cities become divided into sectors to an extent that living conditions of the people who live there become poor. Further, material affluence and consumer habits tend to produce waste to an extent that exceeds available collection, recycling and disposal capacities. City ecology is a new term coined to understand and reverse this development. The term implies that cities should offer their citizens not only water, sanitation, dwellings and transportation but also an environment which includes vegetation, art and facilities for community activity. The human resources are available to deal with these problems but political priorities may cause delays or even cause such environmental restoration projects to be disregarded.

ENERGY AND CLIMATE

Human activities are altering the composition of the atmosphere in such a way that its contents of carbon dioxide, methane, nitrous gases, CFCs (freons) and other so-called "greenhouse gases" are increasing.

The primary effect is a change in the atmosphere's ability to transfer and absorb radiative energy. The physics which link this change to climate changes is quite well understood in principle. Hence, it might be expected that the climatic consequences of anthropogenic pollution should be predictable.

This is not the case yet and may never become reality. The reason is found in the complicated, nonlinear nature of the climate system, which causes the climate to show chaotic and erratic behaviour. It is quite possible - as suggested by The Inter-Governmental Panel on Climate Changes - that increasing concentrations of atmospheric "greenhouse gases" will lead to climatic changes (increasing global mean-temperatures, say) which, in turn, could be followed by severe consequences such as increasing sea levels, draughts and other changes in precipitation patterns.

It cannot be disputed that this is a strong possibility, which deservedly has received much attention while less attention has been aimed at the uncertainty with which the climate and consequence predictions are made. We do not know the sensitivity of the climate to changes of various kinds and cannot, with any certainty, claim that we have seen a clear climate signal as a result of the increase in "greenhouse gases" which we have experienced to date.

The numerical models which are used for climate predictions are inadequate and fraught with faults and short cuts which severely limit their usefulness for decision making. Natural climate variations have, so far, been considerably bigger than the possible man-made ones but, in spite of this, the models have been unable to unequivocally reproduce past climate changes. In short, there is good reason to be prepared to plan with uncertainty.

More than 50 per cent of the total "greenhouse effect" is due to energy conversion; agriculture and changes in land use (e.g. deforestation) account for an estimated 25 per cent while industry (incl. the CFCs) is responsible for approximately 20 per cent of the effect. So part of the remedy will have to be a reduction in the industry's percentage, reforestation, and changes in land use, but more important are measures to slow down and eventually stop the increase in atmospheric "greenhouse gases" which is due to the increasing use of fossil fuels and the resulting emission of carbon dioxide. Fossil fuels quite certainly must continue to be a major component in the global energy household for the foreseeable future. As the demand for higher material living standards - especially in the developing countries - provokes heavier demands on the energy supply, and as growing populations accentuate these demands, we are placed in a situation which may not have any solutions - at least not in the short term. In order to curb population growth, it will be necessary to increase the material living standards, which initially will lead to a reduced death rate and, therefore, to a population increase, and which only can be brought about at the cost of an increase in energy consumption in general, and fossil fuels in particular.

There seems little doubt that the atmosphere's content of "greenhouse gases" will continue to grow even under the strictest of global policies because of the drive in the less developed countries towards higher material living standards. Lower expectations, rather than new technology, are hardly realistic. Furthermore, it should be noted that better housing, refrigerators, etc. cannot be regarded as luxury items but represent means towards better health and less waste. The effects of the drive for better living standards, together with the population growth, are likely to exceed the energy savings which are possible in the industrialized countries. The introduction of primary energy sources free of "greenhouse gases" will be so slow that although the impact of renewable energy sources, nuclear power and energy saving advanced technology will be significant - especially in the long run - it will not prevent the "greenhouse gas" levels from increasing substantially far into the next century; and huge capital investments on a global scale in new technology will be necessary in order just for the carbon dioxide levels in the atmosphere to stop growing so that a new equilibrium can be established.

There are no clear scenarios one may follow and which will lead the world away from the possible dire consequences which may occur. Rather, all possible measures must be taken, including those involved in climate adaptation.

MANAGEMENT OF BIOLOGICAL RESOURCES

The biosphere, of which we are part, can be regarded as a number of ecosystems with the sun as its primary energy source. The ecosystems are interacting systems with a multitude of species playing various roles in the energy consumption and the regulation of processes. In a way these ecosystems can be viewed as an organism which lives as a result of the interplay between its elements.

The ecosystems are being influenced by the fact that 5 billion industrious people live on Earth and multiply, presently by a rate of one more billion over a decade. Increased energy consumption and changes in the climate have been identified as some of the inferences. Man is also manipulating the various ecosystems in a number of other ways. Some problems and options are presented below in order to illustrate the complexity of ecosystem management.

Pest and Pollution Control

Although there is a widespread tendency to see insects as noxious, most are in fact useful, at least as integral parts of nature's equilibrium systems, as only 0.01 per cent cause problems for humans. Typically, these species have reproductive strategies which permit them to show rapid population growth in order to utilize food plants of patchy or temporary occurrence. Such species will show unlimited population growth if man "provides" vast monocultures of their typical food plant.

Agricultural pests are traditionally controlled by chemicals (biocides) which are often used excessively. New and stronger poisons must be introduced constantly, because the insects develop resistance. Many of the chemicals decompose slowly and may be spread by wind and water affecting the food chains and, thus, life in general.

Several studies have shown that the use of predatory insects, mites or microorganisms, and the development of more resistant plant strains represent useful alternatives to biocides. The potential is vast and unexploited. The modest practical use of biological control to date is due to economical pressure for continued use of pesticides. Biocides are also disproportionately cheap in relation to the full social and environmental costs.

Malaria can be caused by a number of *Plasmodium* protozoans and is a major disease in tropical areas. The protozoan is transferred to man, monkey and apes by various types of mosquitos. The density of mosquitos is, thus, a major epidemiological factor in different areas and depends on a multitude of ecological factors: the mosquito, its biotope and the proximity and accessibility to man. The infections may occur sporadically or endemically.

In the 1960's it was confidently assumed that eradication by chemicals would control malaria, however the mosquito's resistance towards DDT proved to invalidate the programmes' success. In addition, the protozoans developed resistance to some of the therapeutic agents, and others, which could be used, are quite toxic to man. Consequently, the infection is endemic, e.g. in vast areas of Africa.

If the mosquito population could be sufficiently reduced, malaria transmission would diminish. However, control will still be difficult as the immunological factors of the infection are very complicated and a satisfactory vaccine has not yet been developed.

For the development of technical methods for pollution control, a number of new techniques are being established but have yet to be fully exploited. Gene technology is particularly remarkable in this connection. By genetic modification it is possible to create microorganisms which in various ways degrade pollutants better than the naturally occurring organisms. Another aspect of new microbe-based technologies is the production of biochemicals that serve as substitutes for the chemical biocides.

Biological Diversity

Most of the world's food production is still based on a dozen species that happened to have been selected for breeding in early human cultures. Recent studies, especially in the tropics, suggest that enormous numbers of good alternatives and supplementary resources exist. However, while the vast majority of living organisms have yet to be tested for usefulness in the material sense, 25-50 per cent of all kinds of living organisms may disappear within the next century unless we manage to stop the destruction of nature.

It is still an illusion to solve this "extinction crisis" with the well-known approach of preserving a selection of "sexy" species, sometimes by captive propagation. In addition, the recent establishment of "banks" of plant seeds and frozen tissue or DNA, holds limited promise. In fact, such approaches may be counterproductive because they divert attention from the real issue by asserting that something is being taken care of.

Biological diversity can only be maintained by preserving samples of undisturbed or still-functional ecosystems. To serve the intended function, however, the selection of sites for preservation must be based on a good understanding of how biological diversity is generated and maintained. Some ecosystems have considerable resilience and the major effort must, therefore, be directed towards areas which can be irreversibly damaged. At least for the tropical forests, identification and protection of centres of endemisms (local concentrations of species of small geographical distribution) may be a useful approach in minimizing the extinction of species. This approach may also be valuable for maintaining stable ecological regions and life support systems in a wider sense.

If we can afford to map the human genome (3×10^9 base-pairs), mapping of a few million plant and animal species should not scare us. For reasons of urgency, however, the selection of top priority sites for conservation actions must be made by explicit analysis over a few groups for which good data are already available.

Conservationists increasingly conflict with both local human populations and with "development" interests. This conflict can only be solved by subdividing the world into areas for man and reserves for biodiversity or the maintenance of essential ecological functions, thereby creating "buffer zones" where these aims can be integrated. In the past, the design of reserves has often been dictated by convenience (lack of conflict) rather than biological criteria. The most cost-effective conservation action would certainly be an explicit analysis of how a maximum number of species can be preserved on a minimum area. Biological knowledge is, therefore, paramount when it comes to the planning and management of grasslands, production forests, wetlands, or the seas' resources. The most crucial problem here is to avoid human activities which irreversibly make once-productive ecosystems less productive or even unproductive.

Ecosystem Management

Recent ecological changes in the temperate and boreal forests is of great concern today. High exploitation and air pollution have lowered the nutritional conditions of many forest districts, and in some areas now literally kill the forest. The interaction between acid rain, wind, soils, leaching of minerals, weathering of rocks, turnover of crops, etc., is complex. The scientific basis for air quality criteria and defining critical loads, e.g. regarding sulphuric and nitrogenous compounds, is still inadequate.

Air pollution must be reduced, but there is also a need to introduce new concepts and thinking concerning forests. Modern forestry should favour indigenous species and ecosystems with variation; and it must be accepted that some areas are set aside to secure biological diversity and ecological functions.

Unlike forests, shallow coastal areas, marshes and waterlogged soils, collectively known as wetlands, do not immediately seem to represent production value. Draining for agriculture and various kinds of misuse has lead to a devastating loss and degradation of wetlands. Unfortunately for society, most of these drainings were profitable only on a very individualistic scale, if at all.

Wetland ecologists have now provided strong evidence that wetlands often represent large economic values, although sometimes in an indirect way. They are often much more productive in their natural state than when converted, and they have important effects in reducing flood peaks, maintaining the ground-water level and stabilizing the climate. Furthermore, they act as sinks of nutrients and contaminants. The loss of these functions is particularly deplorable in the less developed countries, which cannot afford wetland restoration.

Conflicting interests associated with wetlands can be illustrated here by an example from the largest remaining tracts in Europe of natural floodplain forest near Hainburg in the Austrian part of the Danube river. This wetland was protected under the Ramsar Agreement, for preserving recreational values, a special flora, waterbirds, recruitment areas for fish, etc. The wetland also serves as an efficient and cheap waste water treatment plant, giving higher denitrification and better filtering of suspended solids than even highly specialized waste water treatment plants. Preservation of the Hainburg wetlands had to be weighted against the benefits of regulating the Danube for hydroelectric power. Naturally, big economic interests are involved.

ETHICAL, SOCIOECONOMICAL AND POLITICAL STRUCTURES

Environmental problems are characterized by dilemmas and paradoxes behind which are found conflicting values as well as contradictory convictions and conventions. Therefore, to properly assess and maybe solve some of these environmental problems, the ethical, cultural, socioeconomical, ideological, religious and political aspects of the problem need be included.

Dilemmas

The Hainburg (Danube) wetland problem involved a choice between two environmentally attractive options. On the one side production of hydroelectric power, which does not add to the greenhouse effect and (after construction) does not consume natural resources. On the other side reduced pollution of the Danube, conservation of marsh lands and thereby sanctuaries for both flora and fauna. The dilemma is that the wetlands may actually add to the emission of methane. Since the wetland was finally judged to be more valuable than the hydroelectricity and the hydroplant was not built, the necessary electricity has to be provided in other ways. If this implies increased capacity of coalfired power plants, the emission of carbon dioxide may increase significantly.

Nuclear power presents yet another environmental dilemma. Greenhouse gas emission is zero from nuclear energy production. But other environmental risks are evidenced by recent accidents in both East and West, and the problems of safe disposal of nuclear waste still remain unsolved.

The chemical biocides represent another dilemma. On the one side the potential for more food is increased in the form of extra tonnage first harvested and then conserved during storage before consumption. On the other side the potentials and real incidents of pollution are documented. Besides the obvious commercial aspects and regulation problems in a free-trade world, there are other less material issues to consider. E.g. what types of environmental chemotherapy are acceptable? Should sentimental rather than rational arguments decide? And would the dilemma be lessened if the biocides were produced not chemically but biologically, e.g. through processes using gene-modified microorganisms?

There is a need to improve living conditions for people in the developing world. This in turn is sure to increase world population growth - at least for some generations. Which means that in 40-50 years more than a doubling of the present world population will occur. The resulting energy-climate dilemma is gloomy as already described; and it is coupled with a poverty dilemma; i.e. at least in big cities in the developing countries poverty may increase as an immediate result of additional mouths to feed.

The industrialized countries do have quality problems in food and drinking water, as well as needs for improvements in medical care and city ecology. These environmental and health challenges might easily absorb any amount of capital that the industrialized society could make available. At the same time the visible, environmental accomplishments in the industrialized world may satisfy people to an extent that they become complacent.

The developing countries are at the same time struggling for survival at the baseline level of life. How much then, should the industrialized countries invest in basic needs in developing countries instead of in their own environmental improvements? Herein lies a global solidarity dilemma.

In addition to these dilemmas, several paradoxes also characterize the environmental situation.

Paradoxes

Growth of economy is generally measured by the growth of the Gross National Product. However, GNP is not a measure of living standard or life quality, but just a measure of the monetary value of total production in the market sector of the economy. *The GNP paradox* is that the GNP might be reduced by the introduction of cleaner production technologies, which may improve the quality of life and environment.

Several of the Eastern European countries have had a development over the past 45 years with a growing GNP, based on a devastatingly polluting production technology. But depending on the type of raw materials, production technology, packaging, consumer behaviour etc. the same disjunction between nominally improved economy and impoverished environment may be found in other countries as well. There may be good reason to try to establish some kind of positive correlation between measures for wealth, welfare and environmental quality. But presently, there is no paradigm whereby to set such combined goals; and even less so are there measures and means to implement and monitor progressive action.

New technology can be put to work and treat dirty water to any desired quality. Literally speaking, there is no limit as to how much nitrogen, phosphorus or organic matter can be removed in a wastewater treatment plant. So the question is what we consider desirable or enough and whether we are willing to pay the costs. Studies on the receiving waters may even be very inconclusive as to the effect of removing nutrients and organic matter from the waste water. This may be due to the fact that the aquatic ecosystem is too complicated to allow precise predictions of eutrophication or other pollution effects. Or other sources of nutrients, e.g. agricultural land or atmosphere, contribute significantly as well. The technical feasibility and costs for wastewater treatment can be clearly documented, but the effect on the environment cannot. So for the decision-maker, e.g. the politician, the options are not clear. What then happens is that sometimes lack of knowledge is used to avoid or postpone decisions. In other situations decisions are made in spite of rather convincing evidence that no significant effect on the environment can be expected. A *lack of rationality paradox* may be hidden in these examples. But tempting as it is to look for simple and single rationales for making decisions, it is possibly also useless. Decisions are based on political opinion, which is normally multidimensional. Whether decisions are rational, is a question deserving scientific attention. It may be that inter- and multidisciplinary research could reveal hidden rationalities that would increase validity of political judgement and predictability of political decisions, also regarding environmental matters.

Population growth is considered a problem in terms of energy consumption and change of climate. Also food and drinking water supply, sanitation and housing are problems for which alleviation cannot be expected in several years to come. Even with a stable population these problems would be grave and the quality of life poor for many years ahead. Particularly if industrial, materialistic standards are used, the quality of life prospects for large parts of the developing world would be bleak and poor. *The quality of life paradox* is, that the overall improvement of quality of life - including all human, ethical, social and environmental values - is overshadowed by the drive for a higher material standard of living which is only one important quality of life parameter. The key for finding an acceptable global environmental development is not only a change in technology but a change in life-style and consumption patterns in the industrialized part of the world to the effect of reducing the consumption of energy and the pollution of the environment. The developing world also needs to consider the quality of life; but while material wealth must be improved here, without any question, an attempt must also be made to avoid the Western detour, via over-consumption and over-pollution as a consequence of ill-considered industrialization. For the whole world, over-population seems the single, most important threat to the development of an acceptable quality of life for humans and an inherently acceptable change in the environment.

These basic problems have been mentioned in many national and international publications, e.g. the World Commission on Environment and Development: Our Common Future, (1987, Oxford University Press, Oxford); but action is not yet on the political agendas.

CONCLUSIONS AND RECOMMENDED ACTION

The mounting evidence of pollution is to such an extent that it requires immediate action. The evidence is global, e.g. change of climate; or regional, e.g. acid precipitation and chemical pollution of waterways or seas; or local, e.g. lack of sanitation in cities or leaks to groundwater and air from dumping sites.

In order to alleviate these pollution problems and to convert environmental concerns into more concrete action, some high priority tasks are

- to curb the emission of greenhouse gases on a global scale. This demand relates particularly to industrialized countries in several ways; e.g. through reduced energy consumption per capita; more efficient energy utilization; new technologies that specifically prevent the release of greenhouse gases into the outer atmosphere and the stratosphere; and the transfer of capital, technology and know-how to the developing world. The developing world must follow suit as more energy and new technology becomes available to them;

- to reduce the exploitation of both tropical, temperate and boreal forests and conserve, rather than convert, wetlands;

- to change strategy regarding the protection of flora and fauna by applying the concept of natural centres of endemism, rather than continuing futile attempts to save single species, e.g. by relying on conservation through the breeding of animals in captivity or establishing gene banks;

- to develop new concepts and management strategies regarding production, consumption and waste disposal in the broadest sense (gases, liquids, solids). The development of this has already started, cf. terms (and to a certain extent action) such as: life-cycle analyses of products, compounds and elements before starting production; cleaner technologies; wasteless production; treatment with recycling of residues (e.g. phosphorous, nitrogen and sulphur); recycling instead of wasting where appropriate with regard to energy conservation; etc. The necessary, comprehensive planning and implementation approach must be enhanced;

- to increasingly use existing treatment technologies for reducing the emission of acid gases, heavy metals and organic micro-pollutants in order to protect water, air and soil from pollution. Similarly, bans on the use of gases that destroy the ozone layer should be enforced. In principle, the obvious preventive measures already known should be put to work wherever feasible in technical and economical terms;

- to curb global population growth; items such as material wealth and other qualities of life, environmental quality and total population need be addressed more directly, including their interrelationships;

- to significantly increase the capital transfer from the industrialized to the developing world. Over the past decade the capital flow has been away from rather than into the developing nations as a result of repayment of accumulated debts and the payment of interests;

- to develop new strategies for decision-making when dealing with multidimensional problems, uncertainty and even conflicting evidence, not to mention conflicting convictions, values and conventions. This requires better inter- and multidisciplinary research than in the past.

It is a prerequisite for the success of such concrete action that the academic community plays an active role. There seems to be a largely unexploited potential - though no panacea - in the natural (e.g. biological) and technical sciences. This should be mobilized regarding both remedial (clean-up) and preventive action. Possibly a better interaction between pure and applied science holds the promise for less pollution of the environment. But, however promising this may sound, there are barriers that prevent or even prohibit cooperation. These barriers are of an ethical, cultural, socio-economical, ethnic, religious or political nature. Therefore, the humanities, as well as the social and economic sciences must study these barriers and develop new paradigms for how to deal with them in an attempt to secure development and to protect the environment. The international, as well as interdisciplinary, character of this task must be acknowledged.

Public opinion seems very receptive to information, and the news media are more effective than ever before in disseminating the information they select. Therefore, the academic community must recognize its own power and responsibility and apply the same high quality criteria when releasing information to journals or to news media in general.

2.

THE LONG PERSPECTIVE

JØRGEN JENSEN
The National Museum, Frederiksholms Kanal 12,
DK-1220 Copenhagen, Denmark

ABSTRACT

During the last hundred years archaeology has served as an important conceptual and ideological reference to western man. With the development of environmental archaeology which sees the origin of mankind in an ecological perspective, the discipline has acquired a new status because of its profound time perspective and because it allows changes in human existence to stand forth with clarity.

In this article the introduction of farming in Southern Scandinavia is used as a case story. Agriculture is not seen as a simple technological innovation but as a series of new relationships formed between people, land, plants and animals which can only be understood in a holistic perspective. By contrasting the subsistence strategy of the hunters and gatherers of the fifth millennium B.C. to that of the farmers of the fourth millennium B.C. the educational value of environmental archaeology is demonstrated through its description of short time goals versus the perspectives of lasting existence.

Key Words: *Ecology, Environmental Archaeology, Prehistory.*

INTRODUCTION

There are many compelling reasons for studying the origins of mankind when discussing environmental issues. Millions of years have passed since the first man-like creatures appeared on Earth (Leaky, 1977), but *the long perspective* allows changes in human existence to stand forth with clarity. The long perspective also allows us to identify with prehistoric man, because we can see in him a fellow fighter - a creature who, like us, strives for a future.

In the perspective of our long-term existence lies the educational value of archaeology. True, this value has not always been so apparent, because archaeology since its beginnings in the 19th century often has been subject to ethnocentricity. It has been used (and misused), in Europe and in many other areas, to create a national identity. However, in the past few years our views about the origin of mankind have been transformed: archaeology has been influenced by the attention paid to environmental problems. The development of human society in prehistoric times is now often seen as a process in which still larger amounts of energy from the ecosystem are channelled into the cultural system. Of course, no single line of reasoning can give the entire answer to the question: why has human evolution led to the enormous environmental problems which we face today? It can, however, yield some fundamental clues.

It is through this story of humanity in perspective that archaeologists are able to offer some basic concepts for the understanding of man's present situation.

To western man, history has served as the basic conceptual and ideological reference during the last two hundred years, through the period of the formation of the national states: historicism has been the fundamental framework used to understand ourselves as individuals or as nations. We still constantly refer to history at all levels in society, but there is a trend toward a more ecologically and biologically oriented approach. We increasingly refer to "green" rather than historical and cultural values in our society. This is a necessary development, considering our present situation, but it is also necessary to add environmental history as a basic parameter in all environmental decision-making. In nature preservation e.g. it is necessary to include a historical understanding of why we wish to preserve certain types of landscapes instead of others, or why we prefer certain types of balances between plant and animal communities. Even from a purely functionalistic point of view, the historical dimension forms an important pedagogic element in the decision-making, when trying to secure the ecological functioning of, for example, the cultural landscape.

ENVIRONMENTAL ARCHAEOLOGY

In an ecological perspective, archaeology tells about man's ability to manipulate and to simplify nature - an ability which has given him enormous power. If human beings had not been able to manipulate nature, that is, to bring a certain order into nature, the Earth would not have been able to feed more than a fraction of its present population.

This is the intriguing part of the story of the success of the human species; telling how man lived by hunting and gathering during 99 per cent of the time of his presence on Earth. Furthermore, his ability to produce food, enabled the world's population to grow at a rapid rate, from approximately 10 million people 10.000 years ago, to about four thousand million in our time.

The introduction of food production is among the most important achievements of the era of mankind. Agriculture created the economic base and the social environment from which complex societies could emerge. Various hypotheses have been formulated to explain these developments in the Old World. Climatic changes, population growth, demographic disequilibrium and conductive natural habitats are alternately cited as primary causes. Nevertheless, the question as to why agriculture did develop still remains unanswered because, during the later years, archaeology has realized that agriculture is not a simple technological invention. It is rather a series of new relationships formed between people, land, plants, and animals, which can only be understood in a holistic perspective. Cultural systems and their transformation are extremely complex processes. Additionally, the introduction of farming formed the transition to an ecosystem which proved to be fundamentally different from any previously existing ecosystems. Furthermore, agriculture assumed many forms in various parts of the world. Some general trends can be noticed, however: reliance on domestic plants and animals normally began as a minor part of the general subsistence strategy, but the effect of the deviation was to amplify feedback relationships and, thus, this reliance rapidly grew until agriculture was an almost universal means of subsistence. Some of these feedback relationships not only improved the efficiency of agriculture, but made reverting back to semi-sedentary hunting and gathering increasingly difficult (Redman, 1979).

It has therefore become important for archaeologists, in Scandinavia as well, to consider the human community as part of an ecosystem. The system includes the interplay of all organisms with the physical and chemical factors. Yet, it is also important to stress that even if human beings had not existed, nature, or rather the ecosystem, would still be in a continuous process of evolution. In fact, change is an integrated part of the functions of the ecosystem.

This continuous change in the ecosystem was the framework of human existence in Scandinavia throughout prehistory (Jensen, 1982). At the end of the Pleistocene epoch, some 10.000 years ago, dramatic changes in climate and environment occurred as a result of a global warming. When man first arrived in southern Scandinavia, the ecosystem was still altogether young and unstable. Ice still covered parts of Central and Northern Scandinavia, but at the edge of the ice, man lived on the open tundra. All the species which man met in this landscape were what we call opportunists, pioneers. Man himself was equally an opportunist, but practising a rather specialized economy adapted to hunting and the exploitation of a few species - at least during certain seasons of the year, when the small, foraging groups cooperated in reindeer hunting (Grønnow, 1987).

Because of his nomadic way of life, which corresponded perfectly with the drive for rapid dispersion evidenced by other species in the young ecosystem, man made no special impact on the rest of the system. Of course, he did cause some effect as to the lowering of the reindeer population, but if the reindeer population diminished, it only made more room for other animals who competed with the reindeer for space. The archaeological record in Scandinavia gives no evidence for the type of prehistoric overkill which can be seen in other parts of the world. Man living here was still in balance with the surrounding ecosystem.

After man, the pioneer, had spread throughout the immense Southern Scandinavian area, the warmer climate of the postglacial period, however, gradually made the whole scenery change completely. Throughout the millennia the rise of the sea made the land shrink considerably in size, but what the land lost in area, it gained in an overwhelmingly long coastline. The Danish coasts now consisted of deep fjords, sheltered coves, and countless small and large islands. Even if the isolation of the islands in some cases caused a depauperation of the mammalian fauna (Aaris-Sørensen, 1980), the dense climax forest which now covered the land offered a rich variation of food resources, especially in the marine zone. In this landscape man gradually developed into a far more wide-ranging hunter and gatherer who exploited the many possibilities of the productive ecosystem in a seasonal cyclus.

THE VEDBÆK PROJECT - A CASE STORY

During the last 15 years, an important research project, taking place in Denmark at the Vedbæk area north of Copenhagen, has given us the possibility to study the microcosm of such a society of wide-ranging hunters and gatherers, dating from the Atlantic period about 5000 B.C. The society lived at the edge of the climax forest just before farming was introduced in southern Scandinavia (Price and Brinch Petersen, 1987). The productive estuaries, inlets and islands formed by the rising sea were the centre of human settlement during the Atlantic period - and the sea became the main source of subsistence. The area around the Vedbæk inlet is excellent for studying how the well-adapted foraging bands made themselves invulnerable to fluctuations in the available resources through a subtle pattern of settlement.

The richness of the environment at the Vedbæk inlet was reflected in a high density of human settlement. A rich spectrum of food species was available from both sea and land. About 60 species of fish, birds, reptiles, and mammals have been identified at the settlements coming from streams, lakes, wetlands, the inlets, the sea and the forests. Measurements of carbon 13 in human bones from the graves of the prehistoric settlements indicate that the foraging groups depended as heavily on the sea, as did the Eskimos of Greenland from the fifteenth to the eighteenth centuries - about 75 per cent of their diet came from marine resources (Tauber, 1981).

I shall not go into details about how the studies of the spatial arrangement of the sites and the settlements themselves offer substantial clues to the lives and social organization of its occupants. Some of the sites, especially those of the minor islands, were seasonal camps which again were part of a sedentary community onshore. The

Vedbæk project is but one example of the increasing complexity of human society in southern Scandinavia just prior to the introduction of farming; and, the best way to understand it as such, is through an ecological approach.

Aided by their subtle pattern of settlement, the foraging groups ensured that they could exploit the accessible resources at each base camp - though never to such an extent as to exceed the carrying capacity of the area. To secure this restriction, they apparently accommodated the size of their groups and social structure to the ecological situation. In fact, we can see how the changes in the environment were answered by changes in the subsistence pattern of the hunter and gatherer societies.

Only a species in balance with its surroundings can react so sensitively - a species whose major concern is competitive, for example, in a contest for dominance. Through the Vedbæk project we can demonstrate how nature itself develops an incredibly complicated and stable structure to ensure that life has the greatest possible chances of development under all conditions - and that man also follows exactly the same pattern.

FARMING - THE RESULT OF A CRISIS?

It appears that the success of the foraging groups at the Vedbæk inlet, and in Southern Scandinavia as a whole, delayed the introduction of farming for about half a millennium, during which period farming was known, but not adopted here (Zvelebil, 1986). Nevertheless, it is a strange fact that we still only have very vague ideas about what caused the transition from a society of hunters and gatherers to a farming society, a transition which apparently took place within a very short span of time, a few centuries, maybe.

To earlier generations the answer was simple: an invasion of new people occurred, bringing along a new economy and a new technology. Today, with a far more detailed archaeological record, it seems probable that it was the communities of hunters and gatherers themselves that changed from an apparently successful and even comfortable existence to a much more insecure and labour-demanding style of life. But why? A prerequisite for the change was, certainly, knowledge of the new technology.

During the fifth millennium B.C., however, well-established farming communities existed just south of the Baltic, in Central and Northern Germany. Furthermore, the communities of hunters and gatherers of Southern Scandinavia had close contact with these societies and thereby, became acquainted with a new subsistence pattern. There are also indications of a rising population level in Southern Scandinavia throughout the sixth and fifth millennium B.C., i.e. the Atlantic period, which may have caused an unacceptable pressure on food resources. Through an intensification of food procurement, the hunters and gatherers may have compensated for this situation in the beginning, but only to a certain level - that is, up to the carrying capacity of the area. The result may have been a dilemma: either extinction, or change of the basic economy in order to feed the population surplus. This is, however, still a mere hypothesis.

The size of the population in the Atlantic period is very difficult to estimate on the basis of the archaeological records.

On the other hand, there are other factors which may have influenced the situation. As we saw, there are indications that the very productive Atlantic coast region could easily feed even a growing population, but if some of the important food resources disappeared, the situation would have been different. If we picture a society where food resources were exploited to the outmost, the disappearance of only a few of these resources could have far-reaching consequences, especially if it concerned resources which were exploited at a time of the year when a certain general food shortage existed.

This may have been the case with some of the marine food: oysters for example. Oysters were an important resource in the early spring, when most other resources were sparse, but around 4000 B.C. changes in the climate and in the sea level caused a depletion of the oyster beds. The disappearance of this food resource alone did not necessarily have a profound impact upon the societies of hunters and gatherers, but together with other factors, which we don't know so far, it may have caused a crisis that made the introduction of a new economy necessary. It is thought-provoking, nevertheless, that our understanding of man's interplay with his environment is still so incomplete that we cannot explain why the transition to farming took place - and this is the case anywhere in the Old World. Yet such a transition took place. Within a few generations and around 4000 B.C. the whole community changed: husbandry and agriculture became part of the subsistence pattern. Hunting and gathering still continued, but an irreversible process had begun. Man chose a new strategy by which he could further his struggle for dominance. The new strategy meant that man made a deep and lasting encroachment upon the mature ecosystem. This strategy began in south Scandinavia with slash-and-burn agriculture - the burning of the forest - which was a profound attack upon the biomass of the mature ecosystem. New biotopes were created by introducing domesticated animals and plants - and man gradually began to use one of the most effective tools he had ever employed - the plough.

All of this meant that the development of the ecosystem was reversed, back to earlier stages in those areas where the new strategy was introduced. Because those plants that interested human beings were especially the opportunist, or pioneer, ones.

Thus, the subtle mechanisms of population control which the hunter societies had practised disappeared as well. Encroachment upon the ecosystem created the possibility for an as yet unknown rate of growth - but it also created the risk which lies in reversing the development of an ecosystem. Paradoxically, malnutrition and probably famine now became a frequent problem.

With agriculture, man created a young ecosystem in the midst of the dense, well-balanced climax forest. Man now worked against the natural ecological succession. The grain fields of the farmer were the most extreme example of this opposition. Here everything was invested in rapid growth and large production. Every autumn the entire system was destroyed again by the plough.

With agriculture, man lost his place in the ecosystem. The rest of history describes how man gradually destroyed the mature ecosystem, leaving less and less room for those species that naturally belonged to the climax societies. The drama had now begun.

SOME CONSEQUENCES

What happened here, at the northern fringe of the European continent, is only one version of what happened in the rest of the world within a very short period of time, of about 4-5000 years. It is essential to understand what did happen for the sake of our own future as a species.

The hunters and gatherers of the millennia up to about 4000 B.C. had to be finely tuned to their environment in order to survive: they possessed a detailed knowledge about plants and animals. They must have had a deep insight into the resources of their physical world, thus creating an economy which was both extremely efficient and remarkably secure. They lived in small cooperating bands, moving nomadically from camp to camp as the food resources dictated them to do so. They moved across a piece of land which they shared with other bands, probably not devoid of confrontations, but certainly not in search of them as a necessity. Hoarding material possessions must have been foreign to them. They carried their way of life, wherever they went, rather in the form of their intimate knowledge of their environment than in the form of their few possessions.

During the sixth and fifth millennium, however, in Denmark there was a growing tendency to sedentism, and around 4000 B.C., the prehistoric population developed a farming culture that became the exact opposite to that of the hunters and gatherers. Because crops had to be tended and waited for, farmers were obliged to become sedentary, which again offered, for the first time, the possibility of accumulating material possessions. This, in turn, also called forth a whole new aspect of human behaviour. The new economy could support a much greater density of population, giving rise to villages - and towns in certain parts of the world. With the growth of population density, social competition increased, together with the possibility of exercising power over many people on a scale unknown to the hunters and gatherers. The land bearing the crops had to be defended and so had the accumulated possessions. There are clear indications in the archaeological records of growing violence, for example, during this period of time.

There are, of course, many differences in the evolutionary pattern from area to area. One single line of reasoning cannot explain it all, but the basic trend is everywhere the same. Thus the prehistory of Scandinavia, on the fringes of the European continent, becomes just one example of a world-wide phenomenon, illustrating the dramatic breakthrough of a behavioral pattern of mankind, which was not the manifestation of genetically programmed behaviour - but social and psychological consequences of the transition from hunting and gathering to farming.

ENVIRONMENTAL PERSPECTIVES OF ARCHAEOLOGICAL STUDIES

A combined archaeological-ecological approach to the origin of mankind can only to a limited extent contribute to some of the environmental problems which we are facing today. Nevertheless, it has an educational value because of its profound time perspective and because it creates an overall picture of short-term goals versus the perspective of lasting existence. With our focusing upon environmental rescue operations, we often tend to be ahistorical instead of looking into the structural and historical causes of the problems. If we change the balance, we will also see significant changes in the role of history, in the use of the past in the present.

This is why I find it important that we should add environmental archaeology and history to all environmental decision-making. Part of archaeology in Denmark, namely the monument service, has already been merged into The National Agency for Forestry and Nature Conservation. Archaeology and history have something to offer. Historical and archaeological museums are just one element in a pattern where modern tourists, similar to pilgrims, are travelling along fixed routes through landscapes, historical parks, monuments and museums, appropriating knowledge of man's interaction with nature through different historical eras. For generations, traditional archaeology has tried to imitate traditional politico-military history, by replacing statesmen with archaeological cultures and military battles with migrations. Today environmental archaeology has a totally different approach which can contribute to the public's understanding and experiencing of a past with which it is unfamiliar. There are many good reasons for archaeologists and historians to take a keen and critical interest in this new development.

REFERENCES

Aaris-Sørensen, K. (1980), Depauperation of the Mammalian Fauna at the Island of Zealand during the Atlantic Period. Vidensk. Medd. Dansk Naturhist. Foren. 1980, 142: 131-138.

Grønnow, B. (1987), Meiendorf and Stellmoor revisited. An Analysis of Late Palaeolithic Reindeer Exploitation. Acta Archaeologica, Copenhagen, Vol. 56.

Jensen, J. (1982), The Prehistory of Denmark. London: Methuen.

Leaky, Rich. E. and Lewin, R. (1977), Origins. New York. 264 pp.

Price, T.D. and Brinch Petersen, E. (1987), A Mesolithic Camp in Denmark. Scientific American, March, 1987.

Redman, Charles L. (1979), The Rise of Civilisation. From Early Farmers to Urban Society in the Ancient Near East. San Francisco.

Tauber, H. (1981), Kostvaner i forhistorisk tid - belyst ved C-13 målinger.
In: Det skabende menneske. Kulturhistoriske skitser tilegnet P.V. Glob 20. feb. 1981, Copenhagen, Vol. I.

Zvelebil, M. (1986), Postglacial Foraging in the Forests of Europe. Scientific American, May, 1986.

3.

THE EVOLUTION OF THE ENVIRONMENTAL ENGINEER

DANIEL A. OKUN
Kenan Professor of Environmental Engineering, Emeritus
University of North Carolina at Chapel Hill
Chapel Hill, North Carolina 27599, USA

ABSTRACT

Sanitary engineers emerged in the late 19th century with the responsibility for providing safe water and proper sanitation to promote and protect the public health. Their work contributed substantially to the virtual elimination of water-borne infectious diseases, such as cholera and typhoid, in the industrialized world. With the chemical revolution a century later, a new health threat arose from the long-term ingestion of trace concentrations of synthetic organic chemicals in water supplies drawn from polluted sources. Simultaneously, air and soil pollution also threatened the public health with chronic disease. These chemicals were also perceived as degrading the natural environment, threatening all biota. The environmental engineer evolved from the sanitary engineer but was obliged to work with scientists and engineers from a wide range of disciplines.

Meanwhile, water-borne infectious disease is continuing to exact a heavy toll in the developing countries of Asia, Africa and Latin America, where uncontrolled urban growth exacerbates the problems. Hundreds of millions of people in urban areas of developing countries do not have access to safe water or proper sanitation. Despite a considerable investment on the part of external support agencies during the International Drinking Water Supply and Sanitation Decade (1981-1990), the numbers unserved are growing. If environmental engineers are to address this problem with some promise of success, they must work with colleagues in the social sciences.

Key Words: Environment, Public Health, Sanitary Engineer, Urbanization, Water.

INTRODUCTION

This paper, devoted to the environmental edge of my profession, might well begin with the words of the earliest sanitarian:

> *"Thou shalt have a place also without the camp,*
> *whither thou shalt go forth abroad, and*
> *thou shalt have a paddle among thy weapons;*
> *and it shall be when thou sittest down abroad,*
> *thou shalt dig therewith, and thou shalt turn back*
> *and cover that which cometh from thee."*
>
> *(Deuteronomy 23: 12,13).*

Moses was the first "published" sanitary engineer!

The early Romans, too, were sanitarians and engineers. Their grand aqueducts brought water to the cities and their cloacae carried away their wastes. Frontinus (97) introduced a dual water supply system, pure water for drinking and lower quality water for the fountains of Rome and for other nonpotable purposes, a subject to be discussed further.

THE GREAT SANITARY AWAKENING

Despite the early evidence of concern for sanitation, the Dark and Middle Ages were periods of pestilence and plague, marked by a "rift between godliness and cleanliness" (Fair, 1940). Literary witness was given to the circumstances of the period by Samuel Taylor Coleridge after a visit to Cologne near the turn of the century:

> *In Cologne, a town of monks and bones*
> *And pavements fanged with murderous stones,*
> *And rags and hags, and hideous wenches,*
> *I counted two and seventy stenches,*
> *All well defined, and several stinks.*

Ye nymphs that reign o'er sewers and sinks,
The River Rhine it is well known,
Doth wash your city of Cologne;
But tell me, nymphs, what power divine
Shall henceforth wash the River Rhine?

(Coleridge, 1772-1834)

Sanitary crises (we would now call them environmental crises) marked the mid-19th century in England: Cholera epidemics struck London in 1832, 1848-49, 1853-54, and 1865-66; *"India is in revolt and the Thames stinks"* was the report of a distinguished foreign writer (Budd, 1873); 1958 was the "Year of the great stink" (Chapman, 1972); draperies in chambers of the newly-constructed Houses of Parliament on the Thames needed to be soaked in chloride of lime to make them habitable; and there was evidence that Queen Victoria's Consort, Prince Albert, died of mismanaged typhoid in 1861, very probably contracted from drinking Thames water pumped to Buckingham Palace.

In response was what Sir John Simon was to call "The Great Sanitary Awakening." Efforts were initiated to better understand what was happening and why. This is well illustrated by what transpired with the River Thames.

The metropolitan area of London was served in the 1850's by eleven competing private water companies. Most drew their water for distribution without treatment from the tidal reach of the Thames into which the sewers of London discharged. The sewers were originally constructed to carry storm water from the streets of central London directly to the river for the protection of commerce in the growing metropolis. Upon the introduction of water closets the sewers were pressed into service to carry human wastes, as well as the wastes from slaughter houses and other commercial establishments. During dry periods, the river would be low and tides would keep the effluvia pulsing in the river. A floating object was marked to move between London Bridge and Vauxhall for two weeks before it disappeared.

It was commonly accepted at the time, decades before the germ theory of disease had evolved, that the miasmas rising from the river were responsible for disease. Those who were obliged to cross the river eschewed the ferries and even London Bridge, choosing to travel far upstream to avoid exposure to the putrid vapours.

During this period, two of the water companies were competing for customers on the south side of London: The Southwark and Vauxhall Company and the Lambeth Company. The latter chose to improve the aesthetic quality of its product by moving its intake upstream of the heavy pollution.

Dr. John Snow, surgeon to Queen Victoria, had reason to believe, from his epidemiological studies related to the Broad Street well in Soho, that cholera was transmitted by water. He studied the effect of moving Lambeth's water source during the 1854 cholera epidemic in London. He found that, whereas 71 fatal attacks of cholera occurred for each 10,000 homes served by the Southwark and Vauxhall Company, only 5 deaths occurred per 10,000 homes served by the Lambeth Company (Snow,

1955). These findings supported the hypothesis that water was responsible for the cholera epidemic.

An important outcome of this work was that those responsible for providing public water supplies in Europe and the United States sought out waters least subject to contamination, thus establishing the principle of "the best available source."

Towards the end of the 19th century, however, filtration was introduced widely for the treatment of surface waters to improve their aesthetic quality. The introduction of chlorination in the early years of the 20th century assured engineers and municipal officials that almost all polluted waters could be made safe from infectious disease by filtration and chlorination. Accordingly, in the early part of this century the principle of "best available source" was often abandoned in favour of more expedient solutions. Some cities, such as Philadelphia, Cincinnati, New Orleans and London were content to draw water from polluted rivers, comfortable in the knowledge that filtration and chlorination would prevent the transmission of infectious disease. Other cities however, such as New York, Boston, San Francisco, Birmingham and Manchester, developed high quality upland sources, preferring the "best available source," despite higher initial costs.

Attention was then being given not only to avoiding the polluted reaches of the river as a source of drinking water but to eliminating the pollution of the river, particularly where it would impact on the consciousness of the city. Sir Michael Faraday, of electro-magnetism fame, in 1855 wrote to the Times *"If we neglect (pollution of the Thames) we cannot expect to do so with impunity; nor ought we be surprised if, ere many years are over, a season give us sad proof of the folly of our carelessness."* In fact, several commissions had been studying the problem, and a project was proposed by engineer Joseph Bazalgete in 1854 which, after considerable controversy and delay, was finally completed in 1875. This involved the construction of an interceptor along the north bank of the river in concert with construction of the Thames Embankment, one of the great amenities of London. The river near London was protected, but it was not until 1891 that treatment facilities at the outlet of the interceptors were provided to improve water quality in the lower river.

THE SANITARY ENGINEERING PROFESSION

The solution of the problems of water supply and river sanitation became the charge to civil engineers, the "civil" introduced to distinguish them from military engineers. They had become responsible for all types of public works construction: roads, bridges, tunnels, buildings, harbours, etc. Those involved in water supply and sanitation, because they were obliged to know chemistry and biology in addition to construction, formed a specialty among civil engineers called sanitary engineers in the United States, public health engineers in Britain, and hygienistes et techniciens municipaux in France (Claude, 1989). These were the engineers who planned, designed and managed the water supply, sewerage and wastewater treatment facilities so necessary to the functioning of urban society. The virtual elimination of water-borne infectious disease in the industrialized world was testimony to their success.

THE CHEMICAL REVOLUTION

The role of sanitary engineers expanded substantially with the explosion of the chemical industry following World War II. Thousands of new synthetic chemicals, designed to be long lasting and therefore not easily biochemically degraded in the environment, found wide use in serving the needs of rapidly growing populations. Synthetic chemical biocides increased food production and prevented food waste; synthetic fabrics and other materials reduced the cost of household, commercial and industrial products; and new synthetic chemical formulations helped prevent and cure disease. Some 50,000 synthetic organic chemicals are on the market and thousands of new ones are introduced annually.

It was only after long-term use that some of the side effects of the profligate use of these new chemicals, especially the biocides, became apparent. Rachel Carson (1962) in Silent Spring caught the world's attention with her vivid description of the impact of these new synthetic chemicals on biota in the environment. A new breed of epidemiologists, necessarily mathematically sophisticated, made us aware of the long term health effects to humans resulting from long term exposure to even trace amounts of these chemicals in the water, air, and food, and in the workplace. Cancer and heart disease replaced cholera and typhoid in the industrialized world.

Sanitary engineers, who had dealt with the traditional problems arising from biologically-contaminated water, now needed to enlarge the scope of their activities to include the problems of chemically-contaminated air, food, and soil for the sake of the environment as well as to protect human health.

Those cities that chose the expedient approach in obtaining their water supplies, and there are many despite the principle of "best available source," now found that they were facing problems not nearly so tractable as eliminating microorganisms through conventional filtration and chlorination. The chemically-rich waste products of our industrial and urban societies were found to be mutagenic, carcinogenic and teratogenic when ingested over long periods even at trace concentrations.

In an ironic twist, the very disinfectant, chlorine, that was instrumental in eliminating water-borne infectious disease was found to react with natural and anthropogenic organic materials in raw waters to form products that are of relatively high risk to exposed populations. The poorer the water source, the greater the concentration of precursors for the formation of the reaction products, and the higher the disinfectant dosage required. The quality of such waters is more difficult to monitor and the treatment technology is uncertain in its effectiveness. In the United States, the number of contaminants to be monitored under current drinking water regulations will grow from 25 in 1986 to more than 150 by the year 2000. The technical, institutional and financial problems to be faced in providing public water supply are growing more challenging as our knowledge accumulates and as society demands more assurances of safety.

THE ENVIRONMENTAL ENGINEERING PROFESSION

This climate established the need for the environmental engineer; an engineer who is obliged to expand his, or her, knowledge of the sciences and to work with scientists and engineers from a wide range of disciplines. In addition to water supply, myriad other environmental problems have emerged. Addressing the problems of hazardous waste disposal and ameliorating the effects of old hazardous waste disposal sites require contributions from physical, chemical and biological scientists as well as engineers, with the environmental engineer generally playing a key role. At the University of North Carolina, for example, the post-graduate Department of Sanitary Engineering, which had evolved in 1938 from an option in Civil Engineering, became a Department of Environmental Sciences and Engineering in 1963. Students and faculty are drawn from almost all of the basic natural and physical science disciplines as well as from many branches of engineering. The American Academy of Environmental Engineers draws its diplomates from twelve professional organizations.

ENVIRONMENTAL ENGINEERING CHALLENGES FOR THE FUTURE
- The Water Sector

While a wide range of problems relating to clean air, soil, and water and the preservation of the natural resources essential for sustaining life on our planet face environmental engineers, among many other professionals, I will focus on the problem of managing water in the rapidly growing urban areas in both the industrialized and developing world.

In 1950, only two metropolitan areas, New York and London, had populations exceeding 10 million people. In 25 years, by 1975, the number had grown to seven, three in developing countries. After another 25 years, by 2000, it is anticipated that 25 metropolitan areas will exceed 10 million, 20 in developing countries. The United Nations (1986) estimated that by 2010, some 80 per cent, 50 per cent, and 40 per cent of the populations of Latin America, Africa and Asia, respectively, will live in urban areas. The percentages are much higher in the industrialized world.

In these urban areas, which include so-called secondary cities as well as megalopolises, two major issues face those responsible for water management: first, an aging infrastructure in cities in the industrialized world and inadequate infrastructure in the cities in the developing world; and second, inadequate water resources to serve the metropolitan areas, a problem common to both industrialized and developing countries. The matter of infrastructure is primarily a problem of funds. In the industrialized world, painful though it may be to people who are accustomed to being served with safe water 24 hours per day in the home at the turn of a tap, at a very small fraction of a penny per litre, the funds will be found.

Such service is rarely available in the developing world. The health status of populations in the developing world reflect the absence of proper water supply and sanitation. Water-borne infectious disease, virtually eliminated in the industrial world, still take a heavy toll in the developing world, particularly among children. Infant mortality rates, even though understated in Asia, Africa and Latin America, are more than 10-fold greater than in the industrialized world, often exceeding 200 deaths per 1000 live births in the first year. The value of water supply and sanitation in addressing this problem have been well documented (Okun, 1988).

The problems of providing infrastructure for water supply and sanitation in the developing world have occupied the international community since World War II. Progress was slow, so the 1981-1990 decade was declared the International Drinking Water Supply and Sanitation Decade, and the UN family of agencies, including the World Bank, and many other external support agencies, regional, national and non-governmental, were to provide assistance directly and through financial grants and low-interest loans. Despite this intervention, Table 1 shows that, while more people are being provided with services, the urban population unserved with both water and sewerage continues to grow. New efforts are being organized for the post-Decade period, but these too may not reduce the backlog of need.

The matter of obtaining the water resources for urban areas is not a matter of funds alone; political and professional leadership and technical imagination are required. In most countries water for agriculture enjoys a heavy subsidy which permits inefficient use and waste of water in the face of critical urban and industrial needs. In the water-short region of China, which includes Beijing and Tianjin, studies have shown that merely a 10 per cent reduction in agricultural irrigation, without any sacrifice in productivity, would be sufficient to meet the growing urban and industrial needs of the region. The situation is the same in the American west. To manage the resource properly, massive investments are required which in some instances are just not feasible.

A technical approach, water reclamation and reuse, has begun to be adopted in water-short areas. Most urban uses for water do not require water of drinking water quality: industrial processing and cooling, urban irrigation, cleansing, construction, and toilet flushing. Some cities in the industrialized world have long met major needs for industrial cooling water by replacing conventional sources of water with reclaimed water. Dual distribution systems, one system for potable purposes and the other for nonpotable purposes, have been found to be less costly and more politically attractive than investments in new sources of water. One constraint in the developing world is that only small portions of the cities are now served with sewers. With the provision of sewerage in urban areas of the developing world being a high priority for the protection of the public health and with the need for wastewater treatment to prevent serious pollution of surface waters, consideration of water reclamation and reuse may well be appropriate as water needs begin to exceed the resources available (Okun, 1990).

The technical issues facing those responsible for providing water supply and sanitation in the urban area of developing countries are tractable - little new is required. The principal challenge is adapting the technical solutions to the physical, human, and financial resources of the community. The conventional wisdom is that money is the major constraint. However, ample evidence exists that funds are available even in the squatter and otherwise impoverished periurban settlements that characterize urban communities in Asia, Africa and Latin America.

Table 1
Water Supply and Sanitation Services in Developing Countries
1980 - 1990 (population in millions).
(Source: World Health Organization, not yet published).

		1980	%	1990	%	Change	%
Total Population		3236		3991		+755	+23%
Water:	Served	1411	44%	2758	69%	+1348	+95%
	Unserved	1825	56%	1233	31%	-593	32%
Sanitation:	Served	1502	46%	2250	56%	+748	+50%
	Unserved	1734	54%	1741	44%	+7	0%
Urban Population		933		1332		+399	+41%
Water:	Served	721	77%	1088	82%	+368	+51%
	Unserved	212	23%	244	18%	+31	+15%
Sanitation:	Served	641	69%	955	72%	+314	+49%
	Unserved	292	31%	377	28%	+85	+29%

Piped water is generally available to the commercial enterprises and the homes of the upper classes in the city centres. The rapidly growing populations in the periurban areas generally came from villages where women and children carried water, often contaminated, for long distances from streams or ponds. In the cities, they must buy from vendors, paying as much as 30 per cent of their income for water while the well-to-do pay less than 2 per cent (Zaroff and Okun, 1984). Their rates per litre may be 40-fold higher than to those who buy piped water in the same city, and their cost per month for a few 20-litre containers of water per day may often exceed the monthly bills of homes with a kitchen, several baths and a garden. In Lima, Peru, the poor paid three times more per month for buying 23 litres per capita per day from vendors than the rich who used 152 litres per capita per day from the piped system (Andrianzen and Graham, 1974). Were their investments capitalized in piping, they would have better service at less cost than they must bear at present.

The role of the external support agencies should not be primarily in providing money for projects; no matter how much they were to grant or lend, the total would only meet a small part of the need. Their role should be to assist in creating the institutions, including the provision of the necessary human resources, that can sustain the indigenous development of water and sanitation programs. Individual urban projects can be the vehicle for introducing such programs.

If the greater challenges are not technical or financial, but institutional, what is the role of modern environmental engineers? The cities and villages in the developing world are littered with water supply and sanitation facilities that are inoperable or are grossly inefficient. The cause does not rest with poor technology or insufficient funds, but with failed institutions. The external support agencies, and the developing country beneficiaries, have been interested in providing funds for the capital costs of monuments that can be seen and beribboned. Little interest has been demonstrated in preparing a community and its people for the project. The community leaders are often not involved in determining the level of service of the project. They are seldom organized to provide the funds for operation and maintenance of the facilities, let alone capital cost recovery which is necessary if the program is to be sustained. Not having been included in the planning, and often not understanding the benefits to be obtained, they are reluctant to pay. I am pleased to report that international agencies are beginning to appreciate the importance of willingness-to-pay studies, community participation, human resources development and management, all with emphasis on building the capacity to sustain these efforts over the long term.

As environmental engineers have begun to work with scientists and engineers from other disciplines, they now must themselves become fully familiar with the fields of economics and finance, political science and public administration, sociology and anthropology, and education and child care, and must learn to work with professionals from these fields in the promotion, planning and implementation of programs and projects in developing countries. Only through such common enterprise can their programs and projects be sustained.

Just as environmental engineers must reach out to others in the water sector, so they are obliged to join colleagues from other fields and disciplines in executing their responsibilities in other elements of the environment.

REFERENCES

Andrianzen, T.B. and Graham, G.G. (1974), The high cost of being poor - water. Archives of Environmental Health. 28: 312-315.

Budd, W. (1873), Typhoid Fever. London: Longmans and Green; also Arno Press, 1970.

Carson, R. (1962), Silent Spring. Boston: Houghton Mifflin.

Chapman, C.B. (1972), The year of the great stink. The Pharos of Alpha Omega Alpha Honor Medical Society, 35, 3: 90-105.

Claude, V. (1989), Sanitary engineering as a path to town planning; 1900-1920. Planning Perspectives, 4: 153-166.

Fair, G.M. (1940), The genius of sanitation. Journal of the Boston Society of Civil Engineers, April, pp. 67-78.

Faraday, M. (1855), Letter to The Times, July 9, 1955, p.8.

Frontinus, S.J. (97), In "Frontinus and the Water Supply of the City of Rome," translated by C. Herschel, Longmans, Green & Co., New York (1913).

Okun, D.A. (1988), The value of water supply and sanitation in development: An assessment. American Journal of Public Health, 78: 1463-1467.

Okun, D.A. (1989), Issues for the water decade and beyond. Proceedings Water '89 Conference, World Water, Sitra Co., Ltd. Bangkok, KN12-18.

Okun, D.A. (1990), Water reuse in developing countries. Water and Wastewater International, February, pp. 13-21.

Snow, J. (1855), On the mode of communication of cholera, London, John Churchill; also reprinted, Snow on Cholera, New York, Hafner Publishing Co., 1965.

United Nations (1986), Global Report on Human Settlements, Centre for Human Settlements, Nairobi.

Zaroff, B. and Okun, D.A. (1984), Water vending in developing countries. Aqua, 5: 289-295.

4.

THE ENVIRONMENTAL ISSUE AND MEDICINE

BART SANGSTER
Ministry of Welfare, Health and Cultural Affairs
P.O. Box 5406, 2280 HK Rijswijk,
the Netherlands

ABSTRACT

Today's challenge is to restore the balance between man and his environment. The aim is to improve the quality of both man's health and the condition of the environment. The health status and the state of the environment have to be described and understood globally as well as regionally. This implies performing epidemiological surveys and monitoring the environment. Basic scientific work is needed to understand the mechanisms that are involved in causing the present condition of the environment and the state of health. Changes and proposed interventions should always be studied "ecologically", which means that they are to be conceived as the consequence or the cause of disturbances of complex balances. The contribution of medicine in this process is emphasized.

The importance of the proper use of terminology is demonstrated. In order to prevent misunderstandings it is important to be explicit about the level of abstraction at which a problem is discussed, or about the complexity of cause and effect relations which are considered.

Four major determinants: endogenous and exogenous factors, life style and the health care system determine the health condition of a population. This model is considered from an ecological angle and is used to explain the paradoxical situation where the human species seems to be successful whereas the condition of the environment is rapidly deteriorating.

Key Words: Chemical Pollution, Ecology, Environment, Health Care, Life Style, Medicine, Population Growth.

INTRODUCTION

Today, the environmental issue is generally accepted. It is considered of importance to nature, the existence of man and the future of the planet earth. The general public discusses the subject, politicians consider measures and wonder how to make choices, scientists are investigating the extent of the problem and possible solutions. Each issue of a newspaper or a journal contains one or several articles dealing with the present state of the environment. Anybody is convinced that society will have to change in order to cope with the problem, it will change when no solution is found or, when no choices are made, it will change anyhow.

As late as in the 50's, prosperity was depicted by chimneys vomiting large quantities of smoke. Today the same picture is used to symbolize a perishing society. The present interest in the environment originates from concern regarding unwanted health effects caused by pollutants and deleterious effects on the environment per se, as well as upon man's cultural inheritance such as historical buildings.

The question may be asked whether concern for the environment is something new. Already in the prehistoric and historic past, man has been aware of the close relationship between his or her existence and the environment. Man feared changes and the unwanted effects on health. Flooding, drought, animal and plant diseases affecting cattle and crop were very well known major health hazards. Diseases like plague, cholera and typhoid fever were thought to be caused by malignant fumes. Man also tried to influence the environment in order to avert these threats and calamities. The concern of man for his environment is apparently longstanding. However, the nature of man's concern has changed considerably through the ages.

It is interesting to note that within the memory of man, priests were used to carry out the task of influencing the environment. These very same priests also played an important role in curing the individual who fell prey to a disease. Considering the longstanding combination of the role of priest and doctor in one person, it can be concluded that the interest of medicine in the environment probably is as old as mankind itself. In recent history, as well, physicians took a leading role in indicating the environment as a major source of hazardous factors causing the deplorable state of health of the general population in the 19th and early 20th century. These first hygienists were active in the frontline of sanitation of the environment and provided a major impetus in improving the living and working conditions of the population, as well as in providing adequate food and drinking water, to give only a few examples. In those days, physicians informed the general public, politicians and employers about the need to improve the condition of the environment in order to improve health. Today the general public, politicians and employers are demanding attention from physicians to elucidate the relation between the present state of the environment and the possible health hazards.

TERMINOLOGY

Discussions about the relation between human health and the environment tend to be vehement. On the one hand, the threat of pollution is demonstrated passionately whereas, on the other hand, people explain that the adverse effects of environmental pollution are minimal. Obviously, this apparent difference in opinion of the experts adds to the uncertainty of the public. Unfortunately, part of this rather fruitless discussion originates from the improper use of language. A rather trivial example is the frequently used combination of words "environment and health" which usually stands for "environmental pollution and disease". Several key words which often cause misunderstanding will be discussed in this paragraph.

Health: The term "health" can be used in several ways. It may represent the physical and mental condition of one individual or a group of individuals, i.e. the state of health. It may also mean the appreciation of that condition in the terms good or bad, i.e. the quality of health. As already demonstrated in every day practice, health is often used when disease is meant. When it is said that health expenditures are increasing, it is meant that the costs of the ill and the disabled are increasing. Considering the first two interpretations of the term health, it can be concluded, that in the first "health" is used in an objective sense and in the second, in a subjective one. In the first case health, or aspects of the state of health, are to be measured. In the second, a given state of health is appreciated (Figure 1). The implication of the difference will be demonstrated with the following example.

In Western Europe, infant mortality in the 19th century ranged from 100 to 200 deaths per 1,000 live births (Wohl, 1984). Today, infant mortality is about 5 per 1,000. Both 100 to 200 and 5 describe an aspect of health of the population, i.e. state of health. However, 10 deaths per 1,000 in the 19th century would have been described as a health miracle, whereas the same figure today will be considered to be a health disaster. From this example it can be concluded that when discussing "health", particularly in the subjective sense, it is important to indicate what time and what place on earth it refers to.

Environment: In discussions about the relation between environment and health the word "environment" often stands for "environmental pollution", although pollution is only one aspect of the environment. Human health is determined by factors among which many originate from human environment. These factors range from food to air, drinking water and housing, to name only a few.

Pollution: The word "pollution" needs some further consideration. First of all, whether an element in the environment will be called pollution or not depends on time and the place on earth. The example of the smoking chimneys is self-explanatory. Today, pollution is most often associated with chemicals introduced into the environment by man. Since the disaster in Chernobyl, radionuclides are often considered in conjunction with chemicals when using the term "pollution".

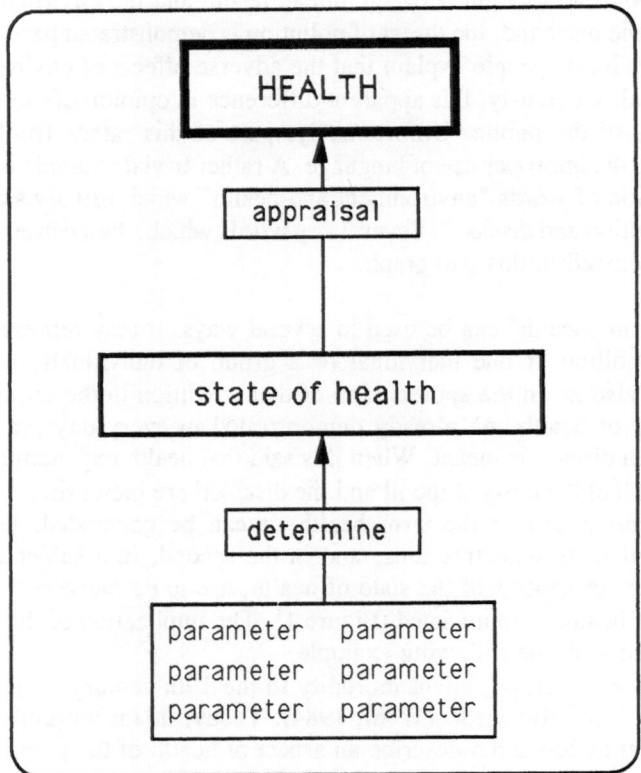

Figure 1. Health and health status.

Only a relatively short while ago, pollution was most often associated with microbiological pollution. When the classical epidemic infectious diseases were a prominent health problem in Western Europe, microbiological pollution of surface water which was used as drinking water and microbial pollution of food, whether vegetable or animal, was a major matter of concern. In other parts of the world, (micro)biological pollution is still one of the most relevant health problems. When discussing pollution, it should be explained whether the term is restricted to man made chemicals, or whether biological and physical pollutants are included as well.

Level of Abstraction: When considering the relation between environment and health, it is worthwhile to determine the type of "relation". Apart from the need to determine the level of abstraction at which the subject will be discussed, it is important to decide the extent of the cause and effect relation one will conceive. Usually, a phenomenon is caused by a chain of causes and effects where each effect is the cause of the next: a cascade of causes and effects. Many of these cascades are occurring simultaneously and are interacting as well. In other words, one has to decide upon the complexity one wishes to take into consideration.

An interesting example is the effect of pesticides, belonging to the group of drins, on birds of prey. At present these persistent chemicals are not used any more in Western Europe, because of the unwanted effects on animals and man. Due to the prohibition of these and other chemicals, after partial extinction, the number of birds of prey is increasing again. However, in areas such as Northern Africa where large swarms of grasshoppers are eliminating all vegetable food for man, livestock and other animals including the birds mentioned, the most effective way to decrease the number of grasshoppers is the use of drins. Notwithstanding the adverse effects of the pesticides on birds, which will be similar in both areas, the use of these compounds will indirectly have a favourable effect on birds of prey in Northern Africa. Due to the availability of feed for prey animals the conditions of the birds of prey will improve. Thus, the net effect of the use of drins on human and animal health in these areas is different from the one in Western Europe.

When people are not aware of the fact that they are discussing the same problem at different levels of aggregation, misunderstanding is the result. In the example, the aggregation level has to do primarily with different processes occurring at the same time. One has to decide whether each process is considered separately, or whether all are studied together. In other words, whether the process will be discussed at the level of individual units in the environment (organism, building), or collections of units (people, town, area), or even higher levels of aggregation, such as countries and fluvial regions, continents and oceans, or at a global level. One has also to decide on the length of time to be taken into consideration when establishing a relation between cause and effect.

DETERMINANTS OF HEALTH

Generally, the health status of an individual organism is considered to be determined by endogenous and exogenous factors. The endogenous factors are represented by the genetic information and the anatomy and physiology associated with the genetic constitution of that particular individual. The exogenous factors are represented by all influences from outside having contact with that particular individual. These environmental factors can be of a biological, physical or chemical nature. This concept applies to all living organisms ranging from viruses to mammals.

Endogenous Factors: Endogenous factors are those properties in the structure and functioning of the organism which are determined by the available genetic information. Obvious properties such as racial characteristics, the colour of the eyes, are endogenously determined. Also differences in biochemistry such as the ability to metabolize ethanol or isoniazid, a drug used in the treatment of tuberculosis, are genetically determined. The same applies to differences in the immune system increasing the susceptibility to infections, such as complement deficiency, or the susceptibility to develop allergic diseases. Thus endogenous factors play an important role in the response, the quality and quantity of the effect, to exogenous factors.

Exogenous Factors: Exogenous biological factors range from viruses to bacteria and parasites that may cause infectious diseases. Examples of physical factors are kinetic energy causing trauma, thermal energy causing burns, ionizing radiation, i.a., causing radiation sickness and mutations, as well as noise causing hearing loss and contributing to the occurrence of hypertension. Chemical factors, whether occurring naturally or man-made, may cause intoxications, mutations and cancer. However, these examples give only unwanted effects of the factors. Exogenous factors may also induce favourable effects. Biological factors, such as bacteria in the intestine, play a role in the defence of the organism towards other biological factors. In man, intestinal bacteria are needed to provide clotting factors. Physical factors, such as thermal energy, are essential for life in providing an adequate ambient temperature. Chemical factors include such essential factors as foodstuffs, water and oxygen.

All exogenous factors of interest have in common that they cause effects (Figure 2). In order to do so, they have to be introduced into the environment. The environment can be conceived to consist of three compartments - soil, water and air - via which the factor may reach man. Indirect routes of exposure are edible plants, which are consumed by man, and more complex routes via the consumption of food from animals.

The circumstances under which the factors are introduced into the environment may vary considerably. This has consequences for the amount of the factor and, therefore, for the effect, both quantitatively and qualitatively. The factor can be generally present. Usually, the level is relatively low and constant. Examples are the generally present pollution of the air with, for example, nitrous oxides, sulphurous oxides, ozone and the ubiquitous presence of polychloro-biphenyls (PCBs). Incidents are connected with localized areas where the factor is present at higher levels than elsewhere. The term incident is associated with environmental incidents such as in Lekkerkerk (the Netherlands) and Love Canal (NY/USA). Two cases which drew considerable public attention in 1980 where inhabitants of a quarter were actually living upon chemical waste (Sangster and Cohen, 1985). Accidents are situations where large amounts of the factor are introduced into the environment in a relatively short period of time in a localized area. Examples range from earthquakes to disasters, such as the ones that occurred in Bhopal and Chernobyl.

In order to cause an effect, the factor has to be absorbed by the individual organism. It can cause local effects, which implies that the mechanism causing the effect occurs where the factor and the individual have contact. The factor can also

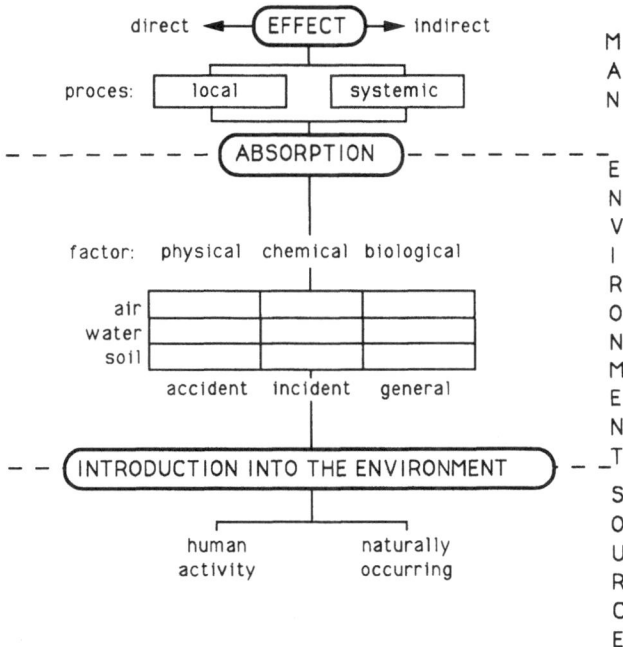

Figure 2. Exogenous factors; chain from cause to effect.

be transported after absorption and cause the effect elsewhere. When chemicals are involved, the nature of the factor can change during transportation because of the biochemical capabilities of the organism (biotransformation). The effect can become manifest immediately after the induction of the process, by an alteration in the functioning of one or more organs causing clinical symptoms. The effect may also take many years to become clinically manifest. This occurs for example in cancer caused by chemicals, ionizing radiation and viruses. These effects are of a more or less direct nature. The factor can also change the response of the individual to other exogenous factors. An example of an indirect effect is the increase of the activity of the enzyme complex P450 in the liver, which increases the toxicity of other chemicals which are depending on biotransformation in order to become toxic.

Life Style: Life style is a third determinant of the health status of an individual or a group of individuals. The term life style applies to eating and drinking patterns, smoking, sexual behaviour, housing and work, to give only a few examples. Differences in life style can cause differences in exposure to exogenous factors, e.g. biological, chemical, physical, as well as psychical factors. However, life style also determines courting and marriage and, therefore, influences endogenous factors too. In closed communities where intermarriage occurs frequently this may induce an increase in inheritable diseases.

An interesting example is the rare genetically determined disease called *porphyria variegata*. The incidence of this disease in South Africa is relatively high. The same applies to a region in the Netherlands around the town of Deventer. It appeared that the disease in South Africa could be traced back to an orphan girl who emigrated from Deventer to Cape Town in 1688. The shortage of women in those days in the Cape, the large number of children and intermarriage among their offspring caused the present high incidence.

Health Care System: The fourth determinant is the health care system. To some extent the quality of the health care system can be conceived as one of the aspects of the life style of society. Prevention, curative medicine and care are distinguished. Primary prevention deals with the exposure to exogenous factors by means of regulation. Thus air quality guidelines are established, maximal allowed concentrations (MAC) for chemicals at the working place and acceptable daily intakes (ADI) for chemicals in food are determined. Other examples are legislation with regard to the quality of food and the chemical and biological quality of drinking water. The aim is to prevent exposure above levels which cause unwanted adverse health effects. Secondary prevention aims at changing the reaction of the organism to exogenous factors. Well known are the paediatric immunization programmes which have greatly reduced the incidence of infectious child diseases such as *tetanus, diphtheria, whooping cough (pertussis)* and *poliomyelitis* in those countries where the programmes were implemented. The worldwide eradication of *small pox* is the best example of the usefulness of this form of prevention.

Curative medicine aims at restoring in those cases where the health condition of an individual or of groups of individuals is affected. Particularly in minor diseases, this aim is often achieved by the patient himself and his or her relatives. In underdeveloped countries this is unfortunately the only possibility to improve health. Nonetheless, curative medicine is for the major part the result of the combined activities of qualified and trained personnel, ranging from general practitioners to medical specialists, nurses, physical therapists, pharmacists and social workers, for example. Curative medicine can be practised in the streets, at home, in the local hospital and in highly qualified hospitals such as university and teaching hospitals.

In those cases where no full restoration of the health condition can be achieved, care is needed to train the patient to participate in social life with his inabilities or handicap, as optimally as possible. Revalidation, physical therapy, social therapy and the use of appliances are needed to achieve these aims. Care can be provided at home and sometimes at the working place. Institutions are needed for those people who are not able to support themselves alone or with the assistance of relatives or friends. Care is also needed for those whose prospects are worse and in whom the condition will gradually deteriorate, such as in many chronic diseases. Sometimes cure and care are called secondary prevention. In that case, the above mentioned primary and secondary prevention both are considered to be primary.

The Net Effect of the Four Determinants Mentioned (Figure 3) determine the health condition of a population at a given moment (Lalonde, 1974). Changes in the health condition reflect changes in these determinants. Careful observation of these changes provides insight into the impact of interventions. Epidemiology in conjunction with many other biomedical disciplines, provides the possibility to study the health condition, changes and the effect of interventions.

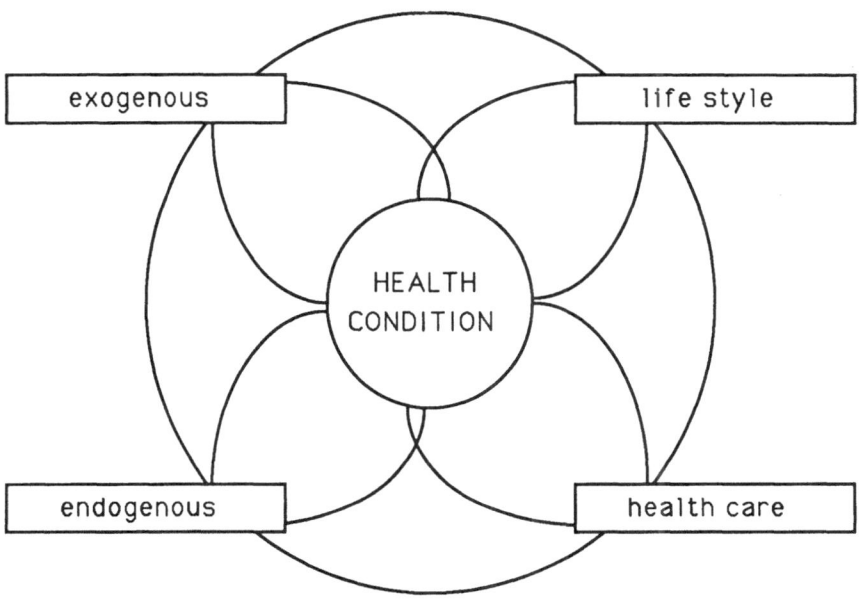

Figure 3. Determinants of health.

HISTORICAL PERSPECTIVE

The basic laws that apply to species in relation to their success apply to man as well. However, man differs from all other species with respect to his ability to influence his environment wittingly. Man uses this ability particularly when his prosperity is threatened by the consequences of these basic laws.

First manifestations of this ability were probably the use of primitive tools in hunting and utensils in preparing food, as well as the use of heat from fire in cooking and the artifice of using hides, for example, as protection from the elements. A major intervention in the natural environment of primitive man was the transition from hunter to herdsman. Ecologically spoken, there had to be a balance between the number of humans in the tribe and the number of animals in the herd in order to survive.

The development into farmer implied a more active changing of the environment e.g. deforestation, ploughing and irrigation. Plant species were selected and plant diseases had to be dealt with. Similarly, the knowledge about animals was used and expanded when man started to raise cattle.

The transition from herdsman to farmer implied that man settled down. Thus villages started to develop and society became more complex. Not all individuals were any longer involved in the primary process of food production. Specialization started. As a consequence, farmers had to produce more than needed by themselves. Part of society became dependent on the success of the farmers, whereas farmers became dependent on the skills of the craftsmen. In order for the society to survive as a whole, the size and needs of the different groups in the society had to be kept in balance.

In a similar way, society in the middle ages can be conceived from an ecological point of view. Towns had to live in harmony with the surrounding villages and hamlets. Farmers improved their capability to produce more than needed by themselves. Part of their products was either traded with the townspeople for goods or taken from them as taxes. In return, they benefited from many activities which were carried out in the towns. These ranged, from protection when aggressors from elsewhere were roaming the country, to the building of dikes, canals and ditches which a.o. protected farmers from flooding, just to give a few examples. The government of the area regulated hunting, clearing of woods, and provided roads and waterways for transportation. Overproduction by farmers was not restricted to food. Also part of their offspring was needed to compensate for the effects of epidemics and excessive mortality in the towns, when compared to the countryside. Further, they were needed to provide the personnel for different duties such as digging, building and soldiering. Again a delicate balance was needed. When a town was imposing too many taxes on the surrounding countryside, the number of farmers decreased due to hunger and associated diseases and, in the end, both were caught in a deflationary spiral (McNeill, 1976).

In the past, a balance had to be kept within relatively small areas. The need for a balance between these areas was limited. Society became, however, more complex in the course of history. As a consequence, the areas in which a balance had to be kept grew from county to country and then to continent. At present, this balance has to be kept at a global scale.

By observing his environment, man tried each time to influence balances and use the natural laws to his advantage. In doing so, the interference with the environment often yielded the desired effect. Due to the complexity of the system, of which man is one of many elements, the effect, however, was most often accompanied by other unfavourable effects which were not always anticipated. Usually, the non-anticipative phenomenon was studied, resulting in a renewed intervention by man, attempting to reduce the unwanted effect and to increase the effect originally aimed at. This caused a continuous process of upsetting balances followed by attempts to restore balances.

Two examples will demonstrate the effect of human intervention causing non-anticipative effects. The first deals with the growing need for energy, which has been exponentially increasing since the industrial revolution. Burning wood and peat could not provide the amount of energy needed by industry. This initiated the use of fossil fuel, consisting of carbon which had been stored for millions of years but originated from the atmosphere. By burning coal and mineral oil, huge quantities of carbon dioxide were released into the environment in a relatively short period of time. This caused a major imbalance in the carbon dioxide equilibrium which did not occur when only nonfossiliferous combustion materials were used, because these contain carbon obtained from carbon dioxide existing in the atmosphere at the moment.

The second example deals with the success of chemistry. This science enabled man to create chemicals which could not be made by natural processes. This ability of man, in conjunction with the growth of chemical industry, caused the introduction of large quantities of chemicals into the environment which were not known to nature. Examples are chlorinated pesticides such as DDT, HCH, HCB, drins, endosulphan, PCP, 2,4-D, 2,4,5-T and PCBs. Other examples are solvents and CFCs. It should be kept in mind that these compounds were developed and introduced into the environment in order to change balances in a direction wanted by man. Unwanted and hazardous effects are usually discovered after the introduction. Highly toxic compounds, such as dioxines, were at first unnoticed as a byproduct, i.a. of herbicides and wood preservatives, and were introduced concomitantly. They are at present recognized as compounds which are generated during the elimination of many products present in waste, particularly when these are incinerated. As a consequence, animals and man are exposed to these compounds. Due to the fact that many of these are able to induce effects in living organisms, including man, the question arises whether adaptation in such a short period of time is possible. Another question is whether the kind of adaptation is favourable for life, or required by man.

During the past centuries, man did not only change his natural environment. He also changed his immediate physical environment. This was partly done in order to protect himself from hazardous environmental influences. Today, in the developed world, humans live in houses which are heated and provided with an adequate supply of drinking water and a sophisticated sewer system. The houses are supplied with furniture and man wears clothes which, to a great extent, are made of synthetic fibres. Man travels in automobiles or uses public transportation systems. Thus, the physical environment has changed considerably. As a consequence, direct contact with the (changed) environment occurs much less frequently than in the past. On the other hand, man is exposed to factors often introduced by himself, in his semiartificial environment.

ENVIRONMENT AND HUMAN HEALTH

Human activities changed man's environment drastically during the centuries. Apart from negative effects on many species originally present, plants as well as animals, this development created many of today's problems, ranging from residues of pesticides in foodstuffs and antibiotics in meat to a part of the acid deposition due to the massive production of manure. However, positive effects were also induced, e.g. sufficient food for the population and, as a side effect, prosperity of some animal species such as the sparrow and the mouse.

        The question, as to what extent human health is affected by the environment(al pollution), may be rephrased to: "to what extent has human health been changed due to the changing environment ?" In answering this question, negative and positive influences have to be considered simultaneously. This should not be conceived as a problem similar to "an equation with two unknowns". Unfortunately, many interrelated unknowns are involved. The problem is complex because man has changed the environment directly or indirectly for the benefit of his own health. Unfortunately, man has not always been conscious of unwanted negative effects on his own health and negative influences on the environment which, in turn, indirectly may affect human health either negatively or positively. All events should be conceived as affecting balances and equilibriums directly and indirectly. In other words, all events, whether natural or initiated by man, should be considered from and ecological point of view.

PARADOX

Evolution has taught that when a particular species manages to live in harmony with its environment, it will prosper. In this context, prosperity may be manifested by an increase in the number of individuals of that particular species. The continuing increase in the number of human beings on this planet, reaching the number of five billion in 1987, and expected to reach the number of 6 billion by the year of 1998, with a birth rate of about 250.000 per day (UN, 1989), at present leads to the conclusion that until now the human species has been extremely successful. This is in contrast to many other species that were less successful and have disappeared, or are about to be extinguished.

        Looking at the human environment, one observes large urban areas, large industrial complexes, severe pollution of the air, rivers and soil, tropical forests that are felled, etc. All these changes of the natural environment, including manipulations in relation to agriculture and stock breeding, are caused by man. The hypothesis that a species is successful only when it manages to live in harmony with its environment seems rather paradoxical, considering the poor condition of the environment, at the one hand and the success of the human species on the other hand (Sangster and van Noort, 1988).

ECOLOGY

As already mentioned, the success of a species depends on its ability to harmonize with its environment. The health condition of a particular species in a particular area can be deduced, i.a., from the number of individuals in that area. When a species prospers, the number of individuals is stable or may even be growing.

The ability to harmonize depends on the balance between the effect of the species on its environment and the effect of the environment on that species. Whether a species can cope depends on the availability of necessary exogenous factors, such as food, and its resistance to hazardous factors, such as viruses and bacteria or extreme temperatures. However, when a species is very prosperous the environment will impose limitations on its growth. A reduction of the quantity of food per capita may cause malnutrition, disease and increase in mortality and even a reduction in the production of offspring. An increase of numbers in a particular area may create a situation which can be successful also to the species pathogenic microorganisms. Then an increase in the population above a critical level may facilitate the occurrence of epidemics, which in turn, reduce the number, unless interventions are carried out - such as vaccination which alters the resistance to the microorganism involved - or hygienic measures, which aim at reducing or abolishing exposure to the agent. These mechanisms apply to all species whether microorganism, parasite, plant or mammal. Ecology is the science that studies the balance and interaction of all species in a particular area.

THE ENVIRONMENTAL ISSUE AND MEDICINE

Medicine taught at university or medical school can be practised as a profession, a science, or can be a university background for individuals involved in a large array of management activities. Medicine may also be the basis for teaching. Physicians can be either practising clinical medicine, or be involved in community medicine or public health. In each of the professional activities, medicine is confronted with the environment and its relevance to health. Usually, the involvement of medicine in environmental issues implies close cooperation with other disciplines.

Physicians, whether being general practitioners or clinical specialists or physicians active in community medicine or municipal health, usually become involved when people are exposed to environmental factors in environmental incidents. This often concerns chemical pollution of soil, air or drinking water. The central question is to what extent has the health of individuals or of groups of individuals been affected, or may become affected, in the future. The challenge is to deal with these problems adequately. Experience is limited, since they were hardly recognized as such before the early eighties. Therefore, training in this aspect of medicine has been started only recently and knowledge is scarce particularly with physicians practising curative medicine.

It is interesting to realize that curative medicine usually deals with the medical problems of individuals. Therefore the approach to the problem in curative medicine differs essentially from the approach in community medicine. In curative medicine the question: "do my complaints originate from pollution?" should be dealt with in a regular, clinical diagnostic way. Apparently, the patient has a medical problem i.e. a complaint or symptom and, therefore, a clinical diagnosis should be made. A history is obtained and a physical examination performed. A differential diagnosis is formulated and additional laboratory tests are done on indication. Usually, this includes toxicological analyses to determine the presence of the pollutants in the patient's body fluids. However, the problem is that also the pollutants have to be considered in formulating the differential diagnosis apart from the regular causes of disease. Limited toxicological knowledge and experience in clinical toxicology may be a problem under these circumstances. In the Netherlands a system has been developed, which anywhere in the country, provides clinical toxicological advice to general practitioners and clinical specialists in polluted areas, and, when needed, assistance in examining patients (Sangster and Cohen, 1985). The system has been in operation since 1982.

At first sight, the question posed to the physicians in charge of the community or municipal health services seems to be similar to that asked to the general practitioners and the specialists. However, their responsibility for the health of the inhabitants of the area involved, primarily concerns the inhabitants as a group and can be effectuated via the local or regional authorities. Therefore, they preferably approach the problem from the side of the pollutants. A health risk evaluation is performed. The results are used by the authorities in deciding how to solve the problem in terms of sanitation, as well as whether additional medical or hygienic measures are to be taken. Also, the question whether an epidemiological study is to be performed or not, can be answered when the health risk evaluation is completed (van den Doel et al., 1988). The results are also important to inform the public about the health impact of the pollution. Usually, performing a health risk evaluation takes a considerable period of time. Close cooperation between curative medicine and community medicine is important because the results of the examination of individuals may provide early signals about the significance of the pollution as far as health is concerned. Similarly, cooperation with veterinary surgeons may be helpful.

In the two situations described, a potential health hazard has been identified and the role of medicine in determining the health impact both at the individual and population level has been described. Curative medicine may contribute in solving the problem by studying individuals in a diagnostic way. Community medicine does so by studying the hazardous factor either by estimating the risk or by performing an epidemiological study in order to determine effects. This approach may be used, irrespective of the nature of the threatening factor. The replica of the situation described occurs when adverse health effects are suspected. This may be when clusters of diseases are noticed and the question is raised whether the cluster is caused by an environmental factor. Then a similar approach may be used in solving the problem. Common to all situations is that medicine together with other disciplines, is engaged in trouble shooting and helping people who may fall victim to the incident.

In close cooperation with many other disciplines, medicine may also be involved in the relation between health and environment in a more academic way. On the one hand, the health condition of a population or subgroup of a population can be determined. On the other hand, the condition of the environment in an area, or aspects of the state of the environment in the area, can be established. Apart from meticulously describing either the state of health or the state of the environment, the question about what causes a given state of health or environment can be studied. This implies a mechanistic approach to the problem. On the one hand, sophisticated epidemiological studies are needed which yield information on the health condition, endogenous factors and exposure data of the population. Fundamental research is needed to investigate biological properties of the exogenous factors whether biological, chemical or physical. Also changes in the health status in conjunction with changes in the (genetic) constitution, changes in exposure, life style and the use and availability of health care, are needed to gain insight in the processes which result in a given state of health. The described incidents and the results of the studies, which are performed when attempting to solve the health related problems in these incidents, may be helpful. They may contain signals regarding the effects and mechanisms involved.

The results of these activities have to be made available to society. Society will then consider its health status. Depending on whether the health status is appraised as healthy or not society will ask for interventions to improve or change its health condition. Public authorities, society and politicians usually discuss and evaluate the state of health and will formulate the changes to be achieved. Civil servants at the ministry of health, officers at the county and municipal health boards, as well as employees of an array of organizations are involved in making the scientific information available and providing information about the feasibility, technically as well as economically, of the improvements desired by society. It is obvious that medicine is involved in this process, as individuals with a medical background are employed by the organizations mentioned.

The process of analyzing the health status and designing and executing intervention strategies is called "public health", which is defined as *the science and art of preventing disease, prolonging life and promoting health through organized efforts of society* (Acheson-Committee, 1988). Being an essential participant in this respectable process is a major challenge to Medicine. The practical implication for those who are medically trained may vary considerably, depending on their occupation.

CHALLENGE

It is generally acknowledged that the present deterioration of the condition of the environment cannot be allowed to continue. In other words, action has to be taken. Considering the large number of problems and the magnitude of each problem, priorities have to be set. Therefore the question has to be asked: "what are the major challenges to man with regard to the present state of the environment and of environmental pollution in particular ?" Because there is a close relationship between the condition of the environment and the health condition of the human species medicine may participate in this discussion and in the decision making, as well.

It is evident, that conclusions of the type: "all pollutants should be eliminated and the environment should be cleaned up completely", are not practical, if not impossible. Following Jean Jaques Rousseau's advice: *"retour ê la nature"*, implicates going back to the number of human beings and their health condition which is associated with a natural, unspoiled condition of the environment i.e. going back several thousand years. Therefore it can be concluded that the question posed should be preceded by decision making. Choices have to be made (WHO, 1985). Afterwards the question can be asked: "how can mankind realize a well-defined quality of the environment and achieve the state of health it prefers?".

When looking at the present state of the environment, the thought may be entertained that the problems are so great that a solution cannot be realized. This fatalistic point of view should be counter-attacked at all levels. It may be worthwhile to realize what efforts are invested, both in terms of human exertion and capital, to create the present infrastructure of society in particular in the industrialized world. Sewer systems and waterworks are constructed. Houses are built. The supply of electricity and gas has been realized by creating extensive and complicated systems. The construction of these facilities has taken many decades, large numbers of people and vast amounts of money. If in the 19th century somebody would have foretold what was about to become reality, nobody would have listened, let alone considered the probability.

Man did not start to build these infrastructures when it became technically possible. Only after it became evident that man might profit from this investment in terms of health and well-being, did he consider these huge efforts. As a consequence, the process of realization was not often initiated by commerce, but by the authorities which in many countries, until the present day, employ these systems. In analogy with what has been achieved in the living environment, it becomes evident that man will only become motivated to go through great pains on behalf of the environment after the benefit from the effort has been made clear. Thus, a major challenge is to provide a realistic analysis of the present condition of the environment and its hazards to human health in order to convince policy makers and the public of the need to do what must be done.

The earlier described ecological approach is a prerequisite for providing the realistic analysis needed. This will be demonstrated with the aid of two examples: overpopulation and chemical pollution.

The extent of the interference of man with the environment is related to the number of individuals and the rate at which the population of the world increases. Therefore, the rate at which the balance between man and environment is being disrupted is increasing, as well. Neither for man nor for the environment benefits can be demonstrated in a continuation of this process. The environmental pollution is associated with the number of individuals. It is also associated with the level of industrialization. As the industrialization of the (third) world is rapidly progressing, the rate of pollution of the world as a whole increases at an even higher rate than the explosion of the world population. There is no doubt that somewhere on the line things will take an unfavourable turn with respect to man and to the environment. Notwithstanding the fact, that it is not known when and where things will go wrong, there is enough evidence for decision-makers to accept their responsibility. The decisions will have to be made and implemented at a global scale. A target will have to be formulated as to the number of humans in this world and the moment when this aim should be met. Different interventions will be needed because situations in different places on earth differ profoundly.

It is interesting to note, that an explosive growth of the population has already taken place in the industrialized world whereas developing countries are in the middle of this process. By the improved health care, especially in the field of prevention, medicine has played an important role in initiating the population growth. A drastic reduction in infant mortality is one result.

Processes that have been effective in the industrialized world to stimulate people to reduce their number of offspring, however, cannot simply be transferred to other countries. As long as people are dependent on their children in order to survive, the population will continue to grow. When primary needs like work, housing, stability in society, health care and commodities such as food, reliable drinking water and energy supplies are available on a regular basis, the need to have large numbers of children will diminish. Only then may life style change and some form of birth control will be practised. Thus, improving the living and working conditions in large parts of the world will be essential to the health condition in other parts of the world and, therefore, to the health status on this planet as a whole. Similarly, efforts aimed at controlling the explosion of the world-population may appear to have the highest priority in terms of preventing a further deterioration of the environment.

The second example concerns chemical pollution. Due to the development of chemistry, information on the exposure of man to chemical pollutants is rapidly growing. The number of compounds that can be detected is ever increasing and the detection limit is rapidly decreasing as well. Even when the level of pollution remains stable this phenomenon by itself creates a picture of a continuously deteriorating environment. This has to be explained to the general public. The advantage of the achievements of analytical chemistry is that the insight into the extent of human exposure becomes more complete.

Many chemicals can be demonstrated in human tissues, blood and urine. With the aid of this information, the question can be answered as to what extent of exposure leads to absorption. Extensive permanent monitoring programmes show changes in exposure and body burden. The question asked by society is: what are the consequences of the exposure to and the presence in the human body of the large array of toxic compounds which are demonstrated? For many compounds, extensive toxicological information is available. The problem is that most information about the toxic effects are collected in animal experiments which are performed to establish safety levels for prevention. It is difficult to extrapolate dose effect relationships from animals to humans. In order to be on the safe side, safety factors are often introduced. However, the results of these types of risk assessments are also used in answering

However, the results of these types of risk assessments are also used in answering questions about the expected effect of a given exposure or body burden. This often leads to a serious overestimation of the expected adverse health effect (Graham et al., 1988). As a consequence, after some time, society will get the impression that the polluting substances are much less hazardous than explained by the experts and will become demotivated in making sacrifices for the environment. Therefore, another way of assessing the health risk has to be employed which provides a more realistic prediction of the effect. This has to be based on animal experimental results, observations in humans and the pathophysiological mechanisms responsible for the effects. As already mentioned, and in general terms, an intervention aiming at improving the health condition of man or the condition of the environment is accompanied by side effects, which are often not anticipated, and which reduce the positive effect while the net effect is usually still positive. Therefore, even the effects predicted by the preferred method will often not become manifest, due to indirect effects associated with the pollutant. The complex interrelationship of effects has to be emphasized each time in order to remain credible.

CONCLUSION

At present, the explosive growth of the world's population is probably one of the most important threats to human health and to the environment. Environmental problems, associated with the production and distribution of food, the use of natural resources such as the rain forests or the production of oil, coal and minerals, and the production of waste ranging from carbon dioxide to dioxins, are quantitatively related to the number of inhabitants of the world. Improving the living and working condition of man on a global scale, are prerequisites for changing this process. In the end, this will affect life style in many respects, ranging from the optimal number of offspring wished and the way man interferes with the environment. However, a stable population is no guarantee for a favourable state of the environment, considering the extent and severity of the pollution in both Western and Eastern Europe and North America.

Both at a global scale and regionally, the health status and the state of the environment has to be described and understood. This implies performing epidemiological surveys and monitoring the environment. Basic scientific work is needed to understand the mechanisms that are involved in causing the present state of the environment and of health. The disciplines needed are not restricted to biomedical and biophysical sciences, but include social sciences. Changes which are observed should always be investigated from two angles: the factor and the effect (WHO, 1988). Changes and proposed interventions should always be studied "ecologically", which means that they are to be considered as the consequence or the cause of disturbing complex balances. Single cause and effect relationships are not to be considered, unless it is kept in mind that these studies are elements in a larger complex system.

REFERENCES

Acheson D, (1988), Public Health in England; the report of the committee of inquiry into the future development of the public health function. p 1. London: Her Majesty's Stationary Office.

Doel, R. van den, Knaap, A.G.A.C., Sangster, B. (1988), Poison Control Centre and Environmental Pollution Health Care and Risk Assessment.
Clinical Toxicology, 26: 89-102.

Graham, J.D. , Green. L.C. , Roberts, M.J. (1988), In search of safety.
Cambridge, Mass. & London: Harvard University Press. pp. 179-219.

Lalonde, M. (1974), A new perspective on the health of the Canadians.
Ottawa: Government of Canada. 73 pp.

McNeill, W.H (1976), Plagues and peoples. New York: Doubleday.

Sangster B., Cohen, H. (1985), Medical aspects of environmental pollution Environmental incidents in the Netherlands 1980-1984. Clinical Toxicology,
23: 365-380.

Sangster B., Noort, R.B.J.C. van (1988), Mens, Gezondheid en Milieu.
Nederlands Tijdschrift voor Geneeskunde, 132: 1101-1105.

UN, (1989), World population prospects 1988. New York: United Nations; population series, 106: 27-42.

WHO, (1985), Targets for health for all. Copenhagen: WHO Regional Office for Europe. pp. 1-201.

WHO, (1988), Priority research for health for all. Copenhagen:
WHO Regional Office for Europe. pp. 9-108.

Wohl, A.S. (1984), Endangered lives; public health in victorian Britain.
London: Methuen & Co. pp. 39-42.

5.

ENVIRONMENTAL PROBLEMS IN FOOD PRODUCTION, PROCESSING AND PRESERVATION

CHAIM H. MANNHEIM
Department of Food Engineering & Biotechnology
Israel Institute of Technology
Technion, Haifa, 32000, Israel

ABSTRACT

The extended use of fertilizers, herbicides, insecticides, etc., has increased crop yields, prevented losses and enabled the provision of a sufficient food supply for the "exploding population" on the one hand, but created serious problems on the other hand. These problems include: undesirable compounds in raw materials; additives used in processing and preservation which may have adverse effects; pollution from processing plants; disposal of food packaging. While much attention has been given to the intentional additives or residues from agro-technological treatments, much less attention has been paid to the potentially harmful compounds which occur naturally in plants used as foods.

In developing countries harvests will be low due to the absence of modern agro-techniques including the use of pesticides, fertilizers, etc., and thus a high level of concern about the lack of sufficient and nutritionally adequate food supply system is inherent. In developed countries the affluence in food supply has not resolved the anxieties associated with it due to fear of potentially harmful residues. The challenge to solve this problem to the satisfaction of all is enormous.

Pollution problems associated with food processing plants are well regulated in most western countries but less so in the eastern countries and in the developing ones. Solutions are costly but available in most cases. The problem of packaging, as a product in the waste stream, has received serious attention only in the last few years. The approaches for dealing with this problem include: reduction of amount of packaging, incineration, pyrolysis, introduction of self-degradable materials, sanitary landfill and recycling. The most promising solutions from the ecological, economic, and sociological points of view are sanitary landfill and recycling. The latter solution still needs considerable research to be universally applicable. Balancing the complex problems of producing and preserving abundant, high quality foods, with the need for strict environmental controls, creates great challenges to food engineers.

Key Words: *Food Packaging; Food Production; Food Production, Environmental Problems in; Preservation; Processing.*

INTRODUCTION

Ecology is defined in the dictionary as the "branch of biology dealing with the habitats and modes of life of living organisms and their relations to their surroundings". The term is derived from the greek word *oikos* i.e. house, and thus ecology can also be "the science of the world seen as man's habitat and his relation to his surroundings or environment".

The so called approaching "ecological disaster" is the disruption of the existing equilibrium between man and his environment. One of the major causes of this disruption is the need to provide sufficient food to the "exploding population" on the one hand without paying sufficient attention to the consequences on the other hand.

The present world population is estimated at around 5,000 million people and it is expected to increase another 20 per cent by the year 2000. Some claim that the world population will stabilize at a level of between 8,000 to 14,000 million in the coming century, i.e. approximately a doubling of the present number. More than 90 per cent of this increase will be in the poor countries and in large overcrowded cities (FAO, 1985).

According to U.N. figures (UNWC, 1987), between 750 and 1,000 million people today are hungry or receive deficient nutrition. On a world wide basis, sufficient basic foods are available; yet their distribution to the needy is often deficient. Furthermore, millions die annually due to consumption of polluted water or microbially contaminated food. Thus, the major challenge to the food scientists and engineers is to provide sufficient food in both quantity and nutritional quality for the rapidly growing world population and, at the same time, to reduce and minimize the environmental problems connected with food production. The environmental problems in food production, processing, handling and distribution include:

1. Undesirable compounds (natural toxicants, pesticides, herbicides, fertilizers, etc.) in raw materials and additives, which may have adverse effects;

2. Liquid, solid, and air pollution from food processing plants;

3. Disposal of food packaging;

4. Excessive use of water and/or energy during food processing, thus depleting natural resources.

POTENTIAL ADVERSE EFFECTS OF FRESH PRODUCTS AND ADDITIVES

The assurance of the safety of the food supply is an essential function of governments. The food industry has an obligation to provide safe, wholesome foods to its customers, but it is the duty of the state to ensure that this obligation is fulfilled.

Foods may have adverse effects which can be traced to compounds appearing in or on them due to agrochemicals, post-harvest treatments, additives, naturally occurring toxicants and environmental contaminants. These include microbial preservatives, antioxidants, stabilizers, colorants, etc., and they are usually well regulated as to their application. For example, Title 21 of the U.S. Food & Drug Administration, (FDA), published annually, describes in great detail the permitted application of intentional additives and specifies maximum doses of use and residual amounts in final product. A list of permitted agrochemicals is given as well as their tolerances in or on the agricultural products which serve as raw materials in the food industry. Furthermore, Title 21 also lists chemicals which may be used as "processing aids" during food processing and which need not be listed as ingredients on the label of the end product. In addition, all packaging materials in contact with foods are subject to very well defined limitations as to composition and possible migrants. Similar laws exist in most other countries, especially in the European Community (EEC, 1989).

Environmental contaminants and pollutants, as compared to intentional additives, are much more difficult to control, especially since they may cross national borders by air, rain or in rivers. Examples in this category include arsenic, cadmium or mercury, which may appear in fish and seafood as well as in many other foods, due to polluted water. Concentration limits for these toxicants are established, by law, in most countries. Polycyclic aromatic hydrocarbons may appear in foods from sources shown in Table 1 and may give rise to levels as those shown in Table 2 (Wren, 1989).

Other examples of pollutants, which may appear in our diet, include chlorinated solvents, PCBs and dioxin. Table 3 shows amounts of chlorinated solvents in drinking water supplies in some European countries, and Table 4 gives data of chlorinated hydrocarbons in human milk. PCBs, which were used in many industries e.g. in printing inks, transformers, cutting oils, etc., have leaked into the environment and have, thus, appeared in foods. Dioxins, while also appearing in nature, were found in bleached paper used in coffee filters, milk cartons etc., in relatively high concentrations. They appeared in milk, packed in cartons, as well as in human milk. Due to an outcry concerning this problem, several years ago, the bleaching process in the paper and board industry has been changed and now papers contain less that a few parts per trillion of dioxins. This is an example of the influence of public opinion in dealing with environmental problems.

While much public concern has been focused on the intentional additives, there has been, in comparison, much less attention paid to potentially harmful chemicals (usually termed "natural toxicants"), such as saponins and glucosinolates, which occur naturally in our diet as a consequence of their presence in plants used as foods.

Table 1
Principal sources of polycyclic aromatic hydrocarbons in food.
From: Wren (1989).

Source	Route	Contaminated Food
Air pollution by combustion	Deposition on growing plant	Leafy vegetables grain fruit
Wood smoke or smoke extract	Smoking of foods	Meat, fish, cheese
Burning fat	Deposition on grilled foods	Charbroiled items
Mineral oils, coal -tar products	Water pollution	Fish, seafood
Mineral oil, wax	Packaging, contact surfaces	Various

In general, the effects of natural toxicants are more likely to be chronic rather than acute. Food is probably the most chemically complex material normally encountered by man. The number of naturally-occurring compounds in food plants exceeds half a million, ranging from volatile flavour compounds to the macromolecular proteins or polysaccharides. Only a minority of these compounds, present in food, have been identified and fewer still have been subject to biological scrutiny at anything like the level carried out on agrochemicals, pollutants, and food additives (Fenwick, 1989). Various reasons may be suggested for this state of affairs, including the cost of pharmacological and toxicological testing and the difficulties encountered in the isolation of pure food ingredients in amounts sufficient for biological examination. It is unfortunate that food scientists have failed to explain to the public that, in the context of food, the word "chemical" refers to both natural and man-made compounds and that, in contrast to what the advertising companies would lead us to believe, "natural" is not necessarily synonymous with wholesomeness and safety.

Table 2
Benzo(a)pyrene (typical levels)
From: Dunn (1982).

Leafy vegetables, greens, fruits	μg/kg		
- Grown in clean air	0	to	1
- Grown in polluted air	1	to	10
Smoked meats, fish			
- Lightly smoked	0	to	1
- Heavily charred	10	to	50
Charbroiled meats, fish			
- Normal cooking	1	to	10
- Heavily charred	10	to	50
Seafood			
- Grown in clean water	0	to	10
- Grown in polluted water	10	to	>100

POTENTIAL FOOD HAZARDS

The six principal categories of food hazards are listed, in rank order, in Table 5 (Fenwick, 1989). This ranking, based upon objective scientific criteria including the severity, incidence and onset of biological symptoms, was determined by Wodicka (1971). The ranking given in Table 5 is based on world wide data, and thus does not necessarily apply to all countries. Nutrition deficiencies are still a major problem in many developing countries. It is estimated that even today about 33 per cent of the world's population suffers from some form of malnutrition (UNWC, 1987). Even in the USA, deaths from nutritional deficiencies (*marasmus* and protein-calorie malnutrition) are more common than deaths from *Salmonella* poisoning (207/million v.s. 102/million people, respectively) (Lee, 1989).

Table 3
Chlorinated solvents in surface and ground waters.
From: Wren (1989).

	Micrograms per litre		
Rivers	Trichloro-ethylene	Perchloro-ethylene	1,1,1-Tri-chloroethane
Sweden (West coast)	0.015	0.005	0.046
England (upstream of industry activity	0.1 to 1.4	to 16	5.41
Netherlands (Rhine)	1.1	1.5	0.2
West Germany (Ruhr at Duisburg)	0.45	0.41	0.05
West Germany (Main and Lippe)	to 70	to 70	--
Switzerland (Lake Zürich)	0.038	0.14	--
Ground Waters			
Switzerland (North)	0.92	0.8	0.27
Switzerland (Zurich)	0.1 to 1.9	0.44 to 2.36	0.04 to 4.8
Italy (Novate Milanese)	7.0	2.8	0.6
Netherlands	to 1100	<22	0.1

Obesity, shortening life-time expectancy, is another serious problem in many developed countries. Hall (1971) has emphasised that the ranking (Table 5) was not linear and that the hazards associated with environmental pollutants and natural toxicants were $\sim 10^3$ less than those originating from nutrient imbalance (excess or deficiency). Contrary to media comment and public concern, the risk due to the presence of pesticide residues or food additives was considered to be $\sim 10^2$ lower than that resulting from natural toxicants.

Hall (1971) described subjective rankings obtained as a result of the comments of the press, food industry, "fringe hysteria" and the FDA regulatory authorities. The only common feature was that natural toxicants were ranked least important by all these groups.

One reason for this situation is the general absence of readily identified, and acute, symptoms associated with natural toxicants. The terminology used may also be confusing; if, in the public mind, "natural" is indicative of goodness, freshness and wholesomeness and encompasses an ecologically-desirable "chemical-free", simple state of living, then "toxicant" is perceived as being related to food poisoning or linked to pollution, adulteration and contamination of the food supply. Thus the term natural toxicant offers an apparent contradiction, describing an area of hazard little understood by consumers and legislators alike. Therefore, Liener (1989) suggested that the term "antinutrient" replace "natural toxicant".

Table 4
Some chlorinated hydrocarbons in human milk.
From: Wren (1989).

	mg per kg Fat		
	PCB	β-HCH	p.p¹-DDE
UK (1982)*	0.50	0.22	1.60
Sweden (1983)*	0.76	0.10	1.05
USA (1983)*	0.97	0.10	1.90
W. Germany (1984)*		2.93	0.211.18
Canada (1986)*	0.65	0.21	0.91

(* Year of publication).

There is no shortage of information available on the biological effects of chemicals isolated from plants, including those consumed by animals and man. The problem is the nature of the available information; the majority of the studies have used in vitro assays, the relevance of which to the human condition is open to question. There is no doubt that the techniques available to the chemist for analyzing food components have generally far outstripped the complementary biological procedures, which are needed to put the analytical data into a meaningful scientific, and social context. This in turn, makes assessment of research priorities difficult for the administrator. The real question is perhaps not "what is the extent of natural toxicant occurrence in food?" (which by and large the chemist can answer) but "what is the *significance* of this occurrence?"

Table 5
Order of priority of actual food hazards.
From : Wodicka (1971).

---

1. Microbiological contamination
2. Nutritional imbalance
3. Environmental contaminant/pollutants
4. Natural toxicants
5. Pesticide residues
6. Food additives

Some examples of natural food toxicants are:

I.   *Glycoalkaloids*. Potatoes and other members of the *Solanaceae* contain toxic glycoalkaloids, mainly alpha-chaconine and alpha-solanine, which interfere with the normal function of the Central Nervous System and also irritate the gastro-intestinal tract (Jadhav et al., 1981).

II.   *Saponins*. Saponins are found in beans, peas and potatoes and their effectiveness in reducing plasma cholesterol levels in animals and man has led to suggestions that their intake should be increased. Such suggestions should, however, be treated with caution, since it is apparent from a variety of uses that saponins also possess potent surface active properties which may affect the permeability of the intestine (Price et al., 1987; Johnson et al., 1986; Gee et al., 1989). This raises the question of interactions between different classes of antinutrients or toxicants.

III.   *Glucosinolates.* Glucosinolates, occurring in leafy plants of the genus *Brassica*, interfere with the thyroid hormone production causing deficiency of this hormone.

The above examples show that any safety policy must embrace considerations of natural toxicants as well as intentional additives, and encompass mechanisms for explaining, simply but concisely, the basis of food safety to consumers. Figure 1 shows a diagram of a natural toxicant research network.

Admittedly, there are considerable socio-economic reasons which will limit the effectiveness of any such program. It is the poorest sections of society which are unable to afford the luxury of implementing dietary and safety recommendations in the selection of their diet - for example, choosing high quality produce, discarding damaged items prior to cooking and avoiding over-indulgence in a limited number of foods. It is, of course, precisely in those sections of society that problems of natural toxicants are inherently greatest. They have the highest ingestion of toxicants and the associated biological effects aggravate existing symptoms of malnutrition.

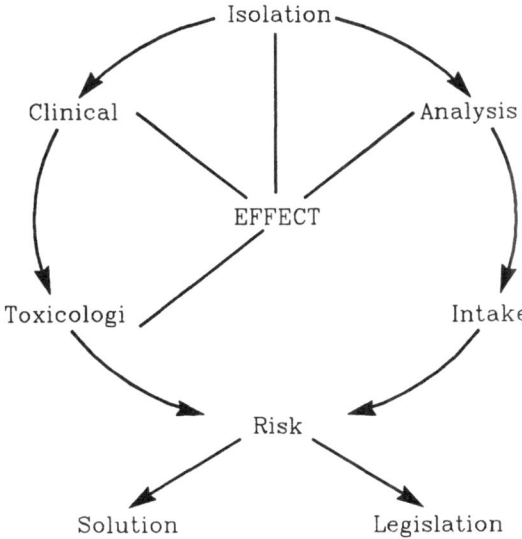

Figure 1. Natural toxicant research. From: Fenwick (1989).

In the less developed countries, where margins of survival are narrow, a high level of concern about the food systems is inherent. The outcome of harvests depends capriciously on the weather, pests, diseases, and on the availability of modern agro-techniques, including the use of better varieties, fertilizers, etc.

Food sufficiency (and surpluses), in the richer nations, was achieved by many years of research to obtain higher yielding and more disease resistant varieties combined with the application of chemical fertilizers, pesticides and herbicides, bringing about the so-called "green revolution". Questions about quality and safety of foods, and pressures from consumer groups, have created new challenges and problems for producers and marketers of food products. Therefore, as part of an overall food safety policy, any evaluation program on potential adverse effects of foods should include *identification and characterization of an effect*, which might be acute or chronic, *methods of analysis* and a *risk evaluation*. Such a study should be followed up by a risk-benefit analysis.

Tasks which regulating agencies and the food and agriculture industries must tackle in future years include (Elton, 1985):

a)   Ensuring that food is as safe as possible, as regards acute and chronic, toxic effects, in the light of existing scientific knowledge;

b)   Improved methods of identifying carcinogens, co-carcinogens and mutagens in in food and quantification of the concentration of such compounds in foods;

c)   Studying the ways in which carcinogens, co-carcinogens and mutagens are formed in foods, and developing methods for reducing their concentrations;

e)   Doing all we can to quantify risks and, if these are appreciable in relation to the other risks of life, take all necessary steps to protect the consumer from these substances in the food supply.

ENVIRONMENTAL PROBLEMS IN FOOD PROCESSING

Environmental problems in food processing include:

a)   Liquid, solid and air pollution
b)   Waste utilization or disposal
c)   Excessive use of energy.

The total amount of water used in food plants varies from about 3 to 4 m^3 to above 10 m^3 per ton of raw material. Problems related to liquid effluents from food plants can be minimized by reducing the total amount of water used in food plants as well as by reducing the amount of organic matter in the waste streams i.e. its biological oxygen demand (BOD). Economizing with water can be achieved by recycling it or reusing it for another step in the operation assuming the pollution degree in the first step is less than in the next one. Heating or cooling systems are examples for operations where almost complete recycling is possible. Water reuse can also be applied in multi step washing operations.

Peeling operations have become less polluting by replacing the chemical lye (NaOH) with physical methods such as abrasive peeling or high pressure steam peeling. Not only do these methods avoid pollution with lye solution but they also save large amounts of water which were needed to remove the lye. These methods also reduce the BOD in the waste stream, and the peel removed in this fashion can be collected and used as animal feed. Regarding solid waste, most plants make every effort possible to sell it as animal feed. One example is the citrus industry where the main form of solid waste are the peels which have become a very good source of animal feed in countries like Brazil and Israel. Other positive examples of waste utilization are the rendering plants in poultry and animal slaughter houses. The entire refuse from these plants is transported, in closed systems, to the rendering plant where the residue is heated to inactivate all potential pathogens, then dried and ground. All these plants are also equipped with special catalytic (or biological) air filters to prevent air pollution.

Great strides have been made by many of the bigger food plants to prevent environmental problems; however, in many of the smaller plants these problems still exist. Unfortunately, the transition to environmentally "clean" plants has not been without casualties. In many instances plants in developed countries, like the USA, were closed, which resulted in a loss of employment, and moved to less developed countries, where there are not as strict environmental laws, thus creating new jobs there.

ENVIRONMENTAL PROBLEMS IN FOOD PACKAGING

The tremendous penetration in the use of plastics in the last 20 years has opened many new opportunities and horizons in all aspects of our life. Simultaneously, however, the widespread use of disposable plastic packages, which replace metal cans and glass containers (both one-way and returnables), started to create serious environmental problems. The need for savings in raw materials of increasing cost, savings in energy to produce plastics, and in the protection of the environment, led various organizations to look for ways by which plastic waste could be reused. Since about 60 per cent of all packaging materials are used to package food products, the food industry, as well as the packaging industry, must address themselves urgently to solve these problems.

It is only when food is consumed that its packaging becomes a waste requiring a management solution. It is a fact that while packaging represents a sta-tistically smaller proportion of litter than is usually realised, its visual effect is disproportionately large (Tuley, 1989).

The composition of home refuse in the USA in 1986 is listed in Table 6 (Morrow et al., 1987).

Table 6
Composition of home garbage in the USA (1986).
From: Ram (1988).

Material	Weight (per cent)
Paper and carton	37.1
Glass	9.7
Metal	9.5
Plastics	7.2
Food waste	8.1
Miscellaneous	28.1

This table shows that, on a weight basis, plastics make up only about 7 per cent of the total home garbage. The different types of plastic materials in the total waste are presented in Table 7. According to this table, as well as data from other countries, more than 50 per cent of all plastic materials are polyethylene (PE) and about 7 per cent are PET - the material which the large carbonated beverage bottles are made of. In the USA about 310,000 tons of HDPE and 290,000 tons of PET were used in 1986. Although these figures may vary from one country to another, the basic composition of plastic waste will be similar (Alter, 1986).

Table 7
Types of plastics in home garbage in USA (1987). From: Ram (1988).

Type of Plastic	Weight (per cent)	
Low density polyethylene	(LDPE)	33
High density polyethylene	(HDPE)	31
Polystyrene	(PS)	11
Polypropylene	(PP)	10
Polyethylene terephtalate	(PET)	7
Polyvinyl chloride	(PVC)	5
Others		3

There are several different approaches for dealing with plastic waste (Pattanakul et al., 1989):

Incineration: The heat produced in this process can be used for power generation or district heating. However, incineration may cause considerable air pollution, requiring special filters or catalytic burners, thus making this process not economically feasible.

Pyrolysis: The decomposition of plastics at elevated temperatures (500-1000°C) can form valuable chemicals, oils and fuel. At present, however, the efficiency of this process is low and its economics are questionable.

Self-Degradable Plastics: Polymeric materials can be made either photo or biodegradable. To be acceptable, a degradable plastic must be stable under normal storage and market conditions, such as exposure to fluorescent light or sunlight. When it does degrade, the end product should have no measurable ill-effects in biological systems. Properties of the plastic should not be altered significantly and the U.S. Food and Drug Administration (FDA) compliance must be assured.

Trials with photo-degradable plastics were disappointing. Under ideal conditions, a dramatic change in the properties of films occurs on relatively short (several weeks) exposure to sunlight. However, the materials do not completely disappear for several months or even years. For thick walled containers, such as beverage bottles, much longer periods, for degradation, are required. Conditions for photo-degrading are usually not ideal and litter may not be exposed to light, because it is buried or lies in the shade.

One of the additives used most, in bio-degradable plastics, is starch. This addition makes the plastic materials much more expensive and changes, adversely, several desired properties including strength and permeability. Another drawback of degradable plastics is that they make in-plant recycling more difficult. During the manufacture of plastics, up to 30 per cent scrap is created, which in most cases is immediately ground up and recycled. Eliminating this part of recycling will increase costs and create further waste problems (Levy, 1987).

Sanitary Landfill: A sanitary landfill is a facility operated under strict rules. Odours, fires, vermin and visual intrusion must be minimized; waste must be compacted; water and gas (methane) evolution must be controlled. Site design must reflect the intended use of the site after it is completed and closed. Landfilling accounts for about 95 per cent of disposal of municipal wastes in the USA, 85 per cent in England, 60-70 per cent in European countries and approximately 30 per cent in Japan (Alter, 1986). Of course, landfilling depends on the availability of suitable sites, thus, this method may be applicable in one country with plenty of empty spaces, like the USA, but unsuitable in crowded countries, like Denmark or the Benelux.

A landfill site may be viewed as a biological reactor in which some of the waste decomposes with time. Time, temperature and moisture content are important. During operation, a landfill is compacted to utilize space and avoid settling. However, as materials degrade and are converted to low molecular liquids, solids, or gases, the landfill settles. Therefore, inert materials must be included to stabilize the site. Plastics are, for all practical purposes, inert in landfills and increase the stability of the site. The methane gas collected, from the landfill, can be used as a source of energy and pay for the operation of the site. Thus, landfilling is an accepted disposal practice for plastics (Alter, 1986).

Recycling: This process, in which collected plastics are either separated into different kinds or reground as a mixture and then reprocessed into new products, is one of the most promising solutions, at this point, to the plastics waste problem (Morrow et al., 1987).

There are 2 basic approaches to removing polymeric materials from the waste stream for recycling. One is to collect separately and recycle selected containers made of a single material, such as PET bottles or HDPE milk containers. However, many waste stream components cannot be recycled efficiently by such separate collection systems and a recycling for mixed plastic streams must be developed.

In the USA, the Plastics Recycling Foundation was founded in 1985 and a Centre for Plastic Recycling Research was established. This Centre works on problems related to the collection, sorting, grinding and reuse of plastic materials. I am not aware of a similar concentrated effort in Europe where each country tries to find its own solution with the main drive being against one-way beverage containers, particularly plastics such as PVC and PET bottles.

About 10 per cent (of 160 million tons a year) of municipal garbage is presently recycled in the USA. Another 10 per cent is incinerated and used for energy, most of the rest is disposed of in landfills. Part of the landfills are also used for energy generation from anaerobic fermentation. The plan is to achieve 25 per cent recycling by 1992, to incinerate about 20 per cent, and leave for landfill only 55 per cent. One should remember that 30 per cent of municipal garbage comes from packages (Table 6). At present, about 20 per cent of paper is being recycled, 40 per cent of aluminum cans and only a small amount of plastic packages. Presently, the main effort in plastic recycling in the USA is being spent on PET and HDPE packages (Pattanakul, 1989).

In order for the recycling of packaging materials to be successful, it is crucial to examine the entire system, from the production of the raw materials (plastics, glass, metal, etc.) through container manufacture, product packaging to recovery, and the reuse of materials. The analysis must be based on total energy requirements and costs as well as on social significance of proposed solutions.

The Industry Council for Packaging and the Environment (INCPEN) in England developed a series of guidelines (Tuley, 1989) to solve the complex packaging-environment problem. These guidelines include:

-       When choosing a package, special considerations as to its total energy requirements throughout the distribution chain, waste arising at all stages of the chain, and the impact of post-consumer waste, must be taken;

-       Avoid the use of excessive packaging;

-       Take positive action to reduce litter;

-       Support resource-efficient reclamation schemes.

CONCLUSIONS

The sources of environmental problems related to food production lie mainly in the efforts to increase yields, prevent food losses and extend storage life of foods. The problems include:

-   Residues in foods, which are derived from chemicals applied in the production of agricultural raw materials with the aim to increase yields;

-   Food additives used to improve shelf life of foods;

-   Liquid and solid effluents from food processing plants;

-   Disposal problems of food packaging materials occurring in municipal wastes.

Unintentional additives in foods, including spray residues and fertilizers, as well as intentional food additives like stabilizers, colours, etc., are well controlled in advanced industrialized countries (AIC) and, to a lesser extent, in less developed countries (LDC). With increasing knowledge about possible deleterious effects, these materials must, however, remain under constant scrutiny.

Efficient waste disposal from food plants is well advanced in AIC, but not so in LDC.

Economic disposal, reuse or recycling of packaging materials still poses a serious problem in all countries and needs considerable, further attention.

REFERENCES

Alter, H. (1986), Disposal and reuse of plastics. In Encyclopedia of Polymer Science & Eng. (Kroschwitz, J.I. ed.) Wiley Interscience Publ., New York, Vol.5: 103-128.

Dunn B.P. (1982), In Carcinogens and Mutagens in the Environment. (H.F.Stich, ed.), CRC Press, Vol.1: 175-183.

EEC. (1989), Council Directive; 89/109. Official J. European Communities, 40: 38-44.

Elton, G.A.H. (1985), Allocation of priorities - Where do the real risks lie? In Food Toxicology - Real or Imaginary Problems? (Gibson, G.G. & Walker, R. eds.) Taylor & Francis Ltd., London.

FAO. (1985), World Food Report. WCED Advisory Panel, Rome.

Fenwick, G.R. (1989), What is the extent of natural toxicant occurrence in food? BNF Nutr. Bul. 14: supplement 1, Oct., 51-63.

Food & Drug Administration (1989), Code of Federal Regulations, Title 21. Superintendent of Documents of U.S. Gov. Printing Office, Washington, D.C.

Gee, J.M., Orice, K.R., Ridout, G.L., Johnson, I.T. & Fenwick, G.R. (1989), The effects of some purified saponins on trans-mural potential difference in mammalian small intestine. In vitro Toxicol. In press.

Hall, R.L. (1971), Information, confidence and sanity in food sciences. Flavour Ind., 2: 455-459.

Jadhav, S.J., Sharma, R.P. & Salunkhe, D.K. (1981), Naturally-occurring toxic alkaloids in foods. CRC Crit. Rev. Toxicol., 11: 21-104.

Johnson, I.T., Gee, J.M., Price, K., Curl, C. & Fenwick, G.R. (1986), Influence of saponins on gut permeability and active transport in vitro. J.Nutr., 116: 2270-2277.

Lee, K. (1989), Food neophobia: Major causes and treatments. Food Technol., 43: (12) 62-72.

Levy, M.H. (1987), A plastics packaging perspective on degradability. Tappi J., 70: (12) 136-137.

Liener, I.E. (1989), Antinutritional factors. In Legumes: Chemistry, Technology and Human Nutrition, (Matthews, R.H. ed.) Marcel Dekker, New York, pp. 339-382.

Morrow, D.R., Amini, M.A., Adams, J.C. & Merriam, C.N. (1987), Overview of plastics recycling. Tappi J., 70: (12) 138-143.

Pattanakul, C., Selke, S., Lai, C. & Miltz, J. (1991), Properties of recycled high density polyethylene from milk bottles. J. Appl. Polymer Sci. In press.

Price, K.R., Johnson, I.T. & Fenwick, G.R. (1987), The chemistry and biological significance of saponins in foods and feedstuffs. CRC Crit. Rev. Food Sci. Nutr., 26: 17-135.

Ram, A. (1988), Ecology and recycling of plastic materials in Israel. Internal Report (Hebrew), Faculty of Chem. Eng. Technion, Haifa.

Tuley, L. (1989), The green response. Food Manuf., 64: (11), 41-44.

Wodicka, V.O. (1971), Quoted in Food Chemical News. 49: 11-17.

Wren, J.J. (1989), Environmental pollution - A threat to food safety? BNF Nutr. Bul. 14: supplement 1, Oct., 27-49.

World Commission on Environment and Development (1987), Our Common Future. Oxford and New York: Oxford University Press, 383 pp.

6.

ENERGY AND ENVIRONMENT: CONFRONTATION OR COMPROMISE?

MARIANO BAUER EPHRUSSI
Programa Universitario de Energía
Universidad Nacional Autonoma de Mexico
Apartado Postal 70-172, 04510 Mexico D.F.

ABSTRACT

Notwithstanding that the need for confronting the environmental impact of our use of energy is recognized by all societies today, the actual individual responses will however, depend on the particular socio-economical, geopolitical and cultural environments. The assignment of priorities cannot be the same for the Advanced Industrialized Countries (AIC) and the Less Developed Countries (LDC). To be constructively successful a global approach must make the necessary allowances for this diversity.

A global energy scenario is first sketched through some quantitative indicators concerning expected economic development, population growth, urbanization trends, and the corresponding envisioned energy needs and environmental impacts. Key contrasts between the industrialized countries and less developed countries are set forth. It is thus shown that the dilemma between energy needs and environmental impacts cannot be stated in absolute terms, valid for all countries. Nor can the standards of the AIC be easily implemented in the LDC which still face problems concerning food, health and low standard of living, or be managed with limited financial resources.

Multidiscipline is the trademark of the energy-environment scenario. Any scientist can find, within his own discipline, many relevant research subjects, both of a basic and of an applied nature, but he can also broaden his scope and join the interdisciplinary effort needed to create viable sce-

narios oriented toward the future. Within the international nature of scientific activities, he can effectively promote concerted and participative actions in the development of technology for protection of the environment. He can also contribute in clarifying the nature and characteristics of the global effort needed on a variety of aspects, to ensure a solution to the conflicting interests as well as the removal of constraints arising from present practices of international financing, trade and technology transfer. In this respect, some relevant suggestions are incorporated.

Key Words: *AIC: Advanced Industrialized Countries, Deforestation, Energy, Energy Efficient Technologies, Energy Intensity, Environment, Global Warming, GNP: Gross National Product, Greenhouse Gases, LDC: Less Developed Countries, Standard of living.*

INTRODUCTION

Energy, or more correctly, energy production and transformation processes are claimed to be the origin of at least fifty per cent of the environmental problems that we face today. Here local and nonlocal effects are included, both in space and time. Indeed, as we are becoming more and more aware of the fact that some of the impacts may be very distant from their source, as much as thousands of miles or hundreds of years. Nevertheless, we bear a responsibility for these with regard to our neighbours and our descendants.

Yet energy is essential for the functioning of our societies. Furthermore, in the near future much more energy will be required world-wide in order to reach a reasonable standard of living for all. Therefore, an energy crisis still lies ahead.

We thus face the question of whether the needed increase in energy consumption can be met without increasing the deterioration of the environment and, at the same time, redressing the damage already done. In the following sections, the framework is sketched in which such questions should be resolved.

THE ENERGY SCENARIO

Energy and standard of living are closely related. Indeed, it is generally accepted that an increased economic development means a higher standard of living equivalent to additional goods and services available to the individual. This, in turn, implies an increased energy consumption per capita. The decoupling between energy intensity (energy per unit product) and gross national product (GNP) that occurred in the seventies since the oil crisis in the high income, advanced industrialized countries (AIC), namely the OECD (Organization for Economic Cooperation and Development), is not present in the less developed countries (LDC) with a middle and low income per capita. This decoupling in the AIC can be attributed, not only to conservation and efficient use measures (Schipper, 1990), but also to a saturation of consumer goods consumption, as well as to a shift in economy towards less energy intensive high technology industries and the service sector. The latter now, in the U.S. for example, occupies over 50 per cent of the population. The energy intensity in the LDC, which on the contrary still have gross deficiencies in their infrastructure concerning housing, transportation, basic heavy industry, etc. is not likely to decrease in the near future, as their energy needs are still considerable, in order to achieve development.

Figure 1 (DOE/IEA, 1990) shows the trends in commercial energy consumption during the last ten years. It can be seen, that the OECD-countries consume as much energy as the rest of the world put together, although they account for only 15 per cent of the population. Note also that the energy consumption has started to grow again in the last three years, breaking away from the decoupling observed in the previous decade, which accounted for a 30 per cent increase in GNP with a practically constant consumption of energy.

Table 1, based mostly on World Bank data and classification (World Bank, 1989), gives more detailed indications on energy, economy and demography, suitable for demonstration of the correlation between development and energy consumption. The countries are grouped into low income, middle income and high income categories, growing from a GNP per capita of $290 to $1,810 to $14,430 calculated in 1987 U.S. dollars. Correspondingly, the energy consumption per capita rises from 297 to 1,077 to 5,000 koe (kilograms of oil equivalent). The top values are heavily influenced by the U.S. and Canada, which carry an average consumption per capita of approximately 8,000 koe while, for Western Europe, it is of approximately 4,800 koe. The U.S.S.R. and other centrally planned European economies which are not members of The World Bank or do not supply data to it (grouped under "WB non-members"), are seen to have energy consumptions comparable to the high income countries, although up to now their standard of living - not easily comparable due to their different economic priorities and political structure - is certainly not equivalent. Their energy intensity is high, similar to that of the middle income countries. The recent political changes will certainly press for a more efficient and clean use of energy, in order to be competitive in a market oriented economy and to allow full integration into the European Community.

The energy data in Figure 1 and Table 1 correspond to primary, commercial energy only. This refers to hydrocarbons (oil and gas), coal and electricity (hydro, nuclear and geothermal). Non-commercial energy (firewood, charcoal, agricultural wastes) is still a significant part of the energy supply in the developing countries (Wood and Baldwin, 1985) and somewhat related to the ratio of rural to urban population, which is seen to decrease from 2.3 for the low income to 0.34 for the high income countries. Generally, non-commercial energy is estimated to contribute about 30 per cent of the energy in the LDC, growing as high as to 90 per cent in Haiti and some African countries. Yet even in partly industrialized countries like India and

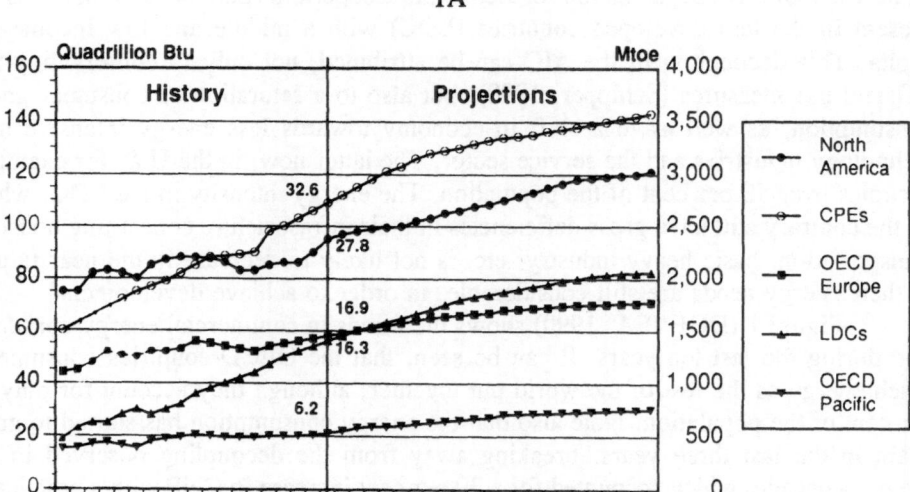

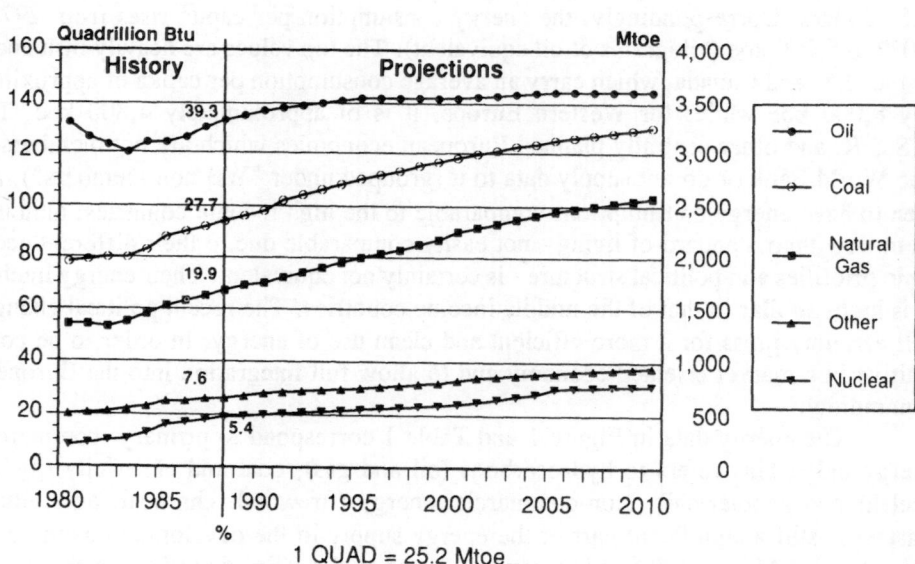

Figure 1. A: World Energy Consumption by region, 1970-2010. B: World Energy Consumption by type, 1980-2010. (Source: U.S. Dept. of Energy, EIA, 1990). (Mtoe = Million tons of oil equivalent).

Table 1

World Commercial Energy Consumption and Standard of Living Indicators (1987). (Source: The World Bank, 1989 -
Note: Figures in parentheses come from different sources.)

	Popu-lation (mill.)	% of world popu-lation	GNP/cap. (1987 U.S. $)	Energy consump. (Mtoe)	% of world energy total	Energy cons./cap. (koe)	Energy intensity (toe/U.S. $ 1000)	Total ext. debt (mill. U.S. $)	% urban popu-lation	Life expec-tancy (years)	Daily Cal/capita (1986)
Low Income Countries	2822.9	56.4	290	856.2	11.2	297	1.045	263344	30	61	2384
- China and India	1866.1	37.3	300	727.8	9.5	390	1.300	76597	33	65	2463
- Others	959.9	19.1	280	128.4	1.7	116	0.477	186747	24	54	2227
Middle Income Countries	1038.5	20.7	1810	1128.5	14.8	1077	0.600	844143	57	65	2855
- Low	609.6	12.2	1200	526.1	6.9	863	0.719	456727	51	64	2777
- High	432.5	8.5	2710	602.4	7.9	1392	0.513	387416	66	67	2970
High Income Countries	774.7	15.4	14430	3845.1	50.5	4953	0.343	30823	77	76	3375
- OECD*	746.1	14.9	14670	3752.4	49.3	5026	0.342	-	77	76	3390
- Others**	30.6	0.5	7880	92.7	1.2	3030	0.384	30823	83	70	3001
WB non-members	365.5	7.4	N.D.	1774.6	23.3	4777	N.D.	N.D.	68	69	3399
- USSR	283.0	5.6	(7340)	(1444.0)	(18.9)	(5102)	0.695	N.D.	67	69	3399
- Others***	82.5	1.8	N.D.	(330.6)	(4.4)	(3735)	N.D.	N.D.	70	71	3400
All countries	5005.1	100.0		7604.4	100.0	1519		N.D.			

* except Greece, Portugal, Turkey, incl. in MIC
** Saudi Arabia, Israel, Singapur, Hong Kong, Kuwait, United Arab Emirates
*** P.R. Korea, G.D.R., Czechoslovakia, Cuba, Bulgaria, Albania, Mongolia

N.D. : No data available
koe : kilograms of oil equivalent
toe : tons of oil equivalent
Mtoe : Million tons of oil equivalent

Brazil, it reaches approximately 48 per cent and 20 per cent, respectively. This has serious ecological impacts. The world-wide trend towards urbanization, especially in the developing countries, tends to alleviate this situation. On the other hand, in those countries it intensifies the demand for commercial fuels.

ENERGY RELATED ENVIRONMENTAL IMPACTS

All energy vectors, from source to end use, have environmental impacts. These arise mainly in the course of the normal operation of the energy system, and are occasionally aggravated by accidental events. The questions which society has to face are diverse. Society clearly has to establish and quantify the environmental impacts connected to normal use. It has to evaluate what levels are acceptable, if any, and whether these can be achieved, and at what cost. Finally, it must be evaluated whether the frequency or character of the possible accidents outweigh the benefits and definitely disqualify any given energy vector.

Energy environmental impacts can be local, regional and global. They arise from various sources: the tapping of energy sources, the transportation of fuels, their transformation and end uses, and, further, the disposal of the final wastes. No detailed description or listing of all impacts can be given here. Illustrations can be found in WEC, 1988 and OECD, 1986. Some gross indications can be obtained from what is stated below, where some of the difficulties of fuel substitution are also pointed out.

Figure 1 also gives the world commercial energy consumption by type. It is seen, that at present 87 per cent comes from fossil fuels (39.3 per cent oil, 19.9 per cent natural gas and 27.7 per cent coal). Their use implies a combustion process that releases - depending on the quality or purity of the fuel and on the completeness of the combustion - sulphur, nitrogen and carbon oxides, particulates, heavy metals as well as unburnt hydrocarbons. The latter giving rise to smog, acid rain and global warming, among other problems. Figure 2 shows the contributions to the greenhouse effect from different sources, with 57 per cent coming from energy production; it also shows the percentages of the different greenhouse gases. Non-commercial energy like firewood is part of the 9 per cent attributed to deforestation, which is also the result of modifications in land use patterns and the commercial demand for timber. Carbon emissions, total and by country, are given in Table 2 (Goldemberg, 1990).

World-wide, up to 62 per cent of the electricity is generated by fossil fuels. The remainder is essentially generated by hydroelectric and nuclear power (19 per cent and 17 per cent, respectively) (Kane and South, 1989). At present, these two technologies seem to be the only alternatives to the fossil fuel combustion process for intensive electricity production. There is still a considerable potential for hydroelectric developments in the world: only 10 per cent of the estimated technical exploitable energy is being used (WEC, 1990). However, the environmental impacts with regard to large dams are both physical and social. There is an increasing awareness and concern about such projects, related to agricultural land and forest flooding, population resettlement, changes in sediment flow, loss of scenic areas and cultural heritage, health impacts, interference with fish migration, downstream problems associated with changed water flows, etc. (Dixon et al., 1989).

The nuclear energy vector is the one vector where there has been a systematic effort, since its inception, to isolate it from the environment, as well as by engineering design as by national and international regulations. However, it faces some severe problems. On the one hand there are still technical aspects like the demonstration of secure waste disposal. On the other hand, there is the problem of how to determine its actual costs, both in absolute terms and also relative to fossil fuels, in view of the drop in oil prices since 1986 and the abundance of coal. Finally, the level of public acceptance is very low - especially since the Chernobyl accident - although not equally low everywhere. Some countries are indeed still committed to strong nuclear programmes, as for instance France, Japan and the U.S.S.R. Concern about the fossil fuel emissions, and the costs to reduce them, as well as design improvements or new technological concepts in nuclear reactors, may bring about a reassessment of the situation (Marques de Souza, 1989).

It is very significant to the environmental question that many AIC effectively reduced their dependency on oil since the oil crisis, giving rise to the drop in oil consumption that can be seen clearly in Figure 1. In addition to nuclear power, oil was partially substituted by natural gas, a cleaner fuel, but also by coal, which is not. Reliance on coal continues to grow, for the geopolitical reasons of averting a renewed OPEC hegemony and because it is the most abundant fossil fuel. It is also geographically more dispersed. At current production rates, coal reserves would last 200 years, as compared to less than 50 years in the case of oil and gas. Projections, like the one shown in Figure 1, point out that the use of coal will eventually equal that of oil, with the price of oil playing a decisive role. Another reason for the rebirth of coal consumption is that its main use lies in the generation of electricity, the demand of which is rising more steeply, world-wide, than that of other secondary energies. This is a trend of development with environmental consequences.

On the positive side, an increase in the use of natural gas, which is envisaged as the fastest growing fossil fuel, is apparent. Considerable reserves exist (3,936 trillion cubic feet, of which 1,500 in the U.S.S.R. and 1,553 in OPEC). However, we cannot expect gas to substitute coal and oil entirely in the generation of electricity and for industrial purposes. This is due to the demands by the residential and commercial sectors for this type of fuel, even though this demand faces the growing penetration of electricity, mainly in the AIC.

Finally, one must consider the transport sector. This sector depends almost exclusively on oil products, i.e., gasoline and diesel - up to almost 100 per cent in most countries. Even with considerable electrified intercity rail systems in Western Europe and many massive city transit systems, in the OECD - in 1980 - there was a 99.2 per cent dependence, with gasoline accounting for 67 per cent (OECD, 1986). The number of vehicles is growing rapidly, both in the AIC and the LDC, because the acquisition of individual transportation is in many ways a status symbol. It is also one of the main contributors to air pollution in the urban areas, up to 80 per cent in the

2A

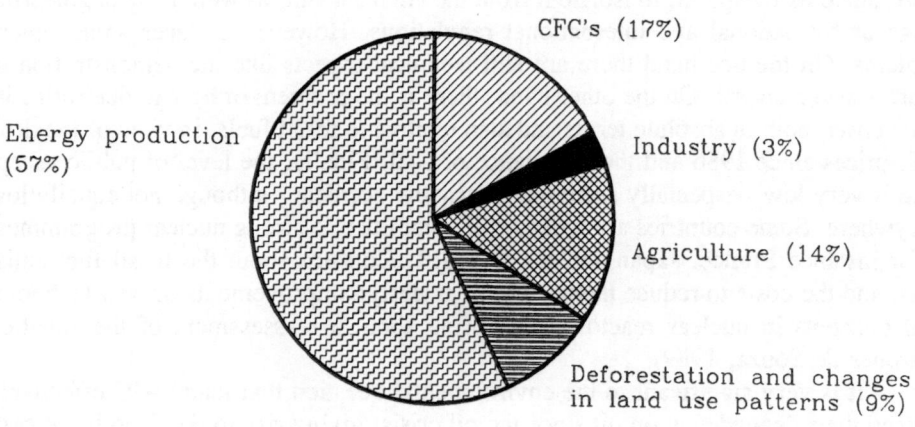

CFC's (17%)

Energy production
(57%)

Industry (3%)

Agriculture (14%)

Deforestation and changes
in land use patterns (9%)

2B

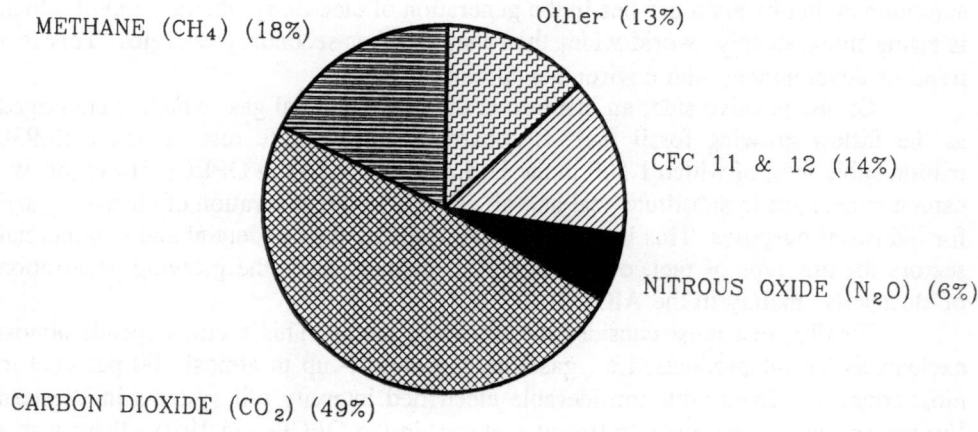

METHANE (CH₄) (18%)

Other (13%)

CFC 11 & 12 (14%)

NITROUS OXIDE (N₂O) (6%)

CARBON DIOXIDE (CO₂) (49%)

Figure 2. Contributions to the "Greenhouse Effect".
(Source, 2A: Goldemberg, 1990. 2B: Kane and South, 1990).

Table 2
Carbon Emissions (CO_2). (Source: Goldemberg, 1990).

	Total (10^6 tons)	%	Per/capita (tons)	From deforestation (10^6 ton)
United States	1201.6	16.9	5.00	-
U.S.S.R	1010.8	14.1	3.59	-
China	554.3	7.7	0.53	-
Brazil	391.5*	5.5	2.83	336
Japan	256.1	3.6	2.11	-
FRG	186.3	2.6	3.07	-
United Kingdom	166.2	2.3	2.94	-
India	144.3	2.0	0.19	33
Colombia	135.8*	1.9	4.68	123
Poland	124.5	1.7	3.32	-
Mexico	106.6*	1.5	1.33	33
Canada	105.2	1.5	4.09	-
France	98.4	1.4	1.79	-
Italy	94.9	1.3	1.65	-
South Africa	92.5	1.3	2.78	-
GDR	92.3	1.3	5.50	-
Czechosl.	65.8	0.9	4.21	-
Australia	61.3	0.8	3.85	-
Rumania	55.8	0.8	2.41	-
Peru	50.8*	0.7	2.56	45
Spain	49.8	0.7	1.28	-
Rep. of Korea	45.4	0.6	1.08	-
Ecuador	43.6*	0.6	2.20	40
All others	1106	15.4		
Def. from all others	1082	15.1		
World total	7169.8	100		

* Deforestation included

Metropolitan Area of Mexico City, for example. It is estimated, that about 14 per cent of the total world-wide release of carbon is due to fuels used in transportation. Also released are hydrocarbons and nitrous oxides. Yet, at present, there is little prospect of any substantial fuel substitution in this area, except for other liquid fuels such as alcohol or methanol which carry their own environmental problems.

Generally, renewable energy sources are recognized as being less detrimental to the environment than fossil fuels, although a clear evaluation should consider the complete system, including plant construction materials. To this effect, a comparison of the CO_2-emissions from electricity generation technologies has been made (San Martin, 1989). It is noticeable that the wood combustion process producing high emissions of CO_2 can be counterbalanced if coupled with fuel regrowth (forest management).

However, the renewable energy sources, other than large hydroelectric plants, will still take some time before becoming significant in the restructuring of the world's energy system. At present they contribute 0.1 per cent of the global energy supply - a major part being geothermal energy. Projections made by the International Energy Agency predict a maximum of 5 per cent in the AIC for the year 2010. The World Energy Conference estimates only 3 per cent by the year 2020. Major technological advances are still needed to consider the large scale energy production as viable (Hartley and Schueler, 1990), although important commercial demonstration projects exist already, or are being planned. Land requirements can be a problem in solar and aeolic applications and, even more so in the use of biomass, when energy crops must compete with food crops for land.

Concern about energy related environmental impacts has been growing in the last decade (WCED, 1987; OECD, 1983 a, b, 1985, 1986, 1987; WEC, 1988). The size and diversity of the task ahead and the problems connected to it, are considerable. According to the U.S. Environmental Protection Agency for example, the man-made emissions must achieve the following reductions to stabilize the atmospheric concentrations at the present levels: 50 per cent in carbon dioxide (CO_2), 10 per cent in methane (CH_4), 75 per cent in clorofluorcarbons (CFCs) and 80 per cent in nitrous oxide (N_2O). Slightly larger figures appear in the recent Intergovernmental Panel on Climate Change report. These figures leave behind the more moderate goal of a 20 per cent reduction in greenhouse gases by the year 2025 established by the Toronto Conference on Climate Change in 1988.

Diverse efforts are programmed at present in most of the AIC and some of the LDC. They should be contrasted with the following. Power generation is one of the main contributors to unwanted emissions. The Electric Power Research Institute estimates the capital cost of CO_2 control for U.S. power plants alone, to be U.S. $584,000 million or U.S. $1,230 per kW (Kane and South, 1989). Even with lower cost estimates, the AIC faces some difficult economic and political decisions, like the creation of special taxes. These are never easily accepted at a national level, unless they are considered to be fairly distributed and their purpose clearly established. A tax on gasoline in the U.S. comparable to that levied in other OECD countries, would have a considerable impact both as a restraining measure and as a source of funds for environmental protection. It will however face very strong opposition.

Also, unilateral action taken by any one country, or merely more stringent limits on emissions, can put its industry at a commercial disadvantage with respect to other countries (Schmidt, 1989) by raising its production costs. An OECD study estimates that environmental control capital costs in a coal-fired plant can range from 5 to 36 per cent of the total capital expenditure, depending on the type of coal and the stringency of the emission limits. The operating costs are also increased (OECD, 1983b).

The impact of carbon emission restrictions on the economy has been modelled (Manne and Richels, 1990), taking into account projected growth rates as well as the cost of possible fuel substitutions and technological developments. A scenario in which the AIC stabilize their carbon emissions at the 1990 level by the year 2000, reducing them afterwards by 20 per cent by the year 2020, and in which the LDC are required to limit their emissions to twice their 1990 levels, will yield the following annual losses - given as a percentage of the aggregate GNP by the year 2100: 2 per cent for the U.S., 1.5 per cent for the other OECD countries, 3 per cent for the Soviet Union and Eastern Europe, 3 per cent for the LDC, with the exception of China, which reaches 10 per cent due to its dependence on coal.

Here we face the dilemma of the LDC. To bring the LDC to a level of commercial energy consumption of about 3,000 koe per capita, that is, the low end of the AIC range, would signify a world consumption of 2.5 times the present amount. The middle income countries would require a continued 7 per cent annual growth for 15 years to achieve this, whereas it would take over 30 years, at the same rate, for the low income countries. Such lapses are increased if one takes into account the expected population growth, still at a 2 to 3 per cent annual rate for the LDC, and more if the above economic impact of carbon restraints is included.

No such GNP growth is contemplated in any of the comfortable scenarios, as far as energy availability for the AIC in the coming decades is concerned - like those shown in Figure 1. These are based on an average assumed growth of about 2.5 per cent for the period up to the year 2010 which is hardly above the growth of the population. The problem of foreign debts (Table 1), if unresolved, may frustrate even such low economic growth. In 1989, the foreign debt of the LDC was estimated to be 1,300,000 million U.S. dollars. This amounted to fifty per cent of their combined GNP and between two and three times their export of goods and services. Furthermore, there has been a net outflow of capital in the last six years. Even with a low growth rate, the LDC will however increase their consumption of energy and will thus face a dilemma concerning the associated environmental problems. The LDC are aware of and concerned about pollution and the impact on the ecosystems in their own cities and countryside, but the resources needed to prevent these impacts compete with other urgent needs. The health and food problems of many LDC, - somewhat hidden in the averages shown in Table I, as far as life expectancy and daily calories per capita is concerned - could soon become critical. The trend toward urbanization in the LDC - where the most populated cities of the world will be predominantly located in the near future - is more the result of rural impoverishment than of the availability of jobs in industry and the service sector. The growth of slums is already giving rise to many vexing "social pollution" problems in need of urgent attention.

The question of global impacts leads to a different discussion. To begin with, the LDC may not accept equal responsibility, in view of the unequal access to commercial fuels in relation to the AIC. They may not be willing, or able, to implement clean fuel technologies, due to their additional cost. Yet, the participation of the LDC can make a dramatic difference in the future, as can be seen in Figure 3 (Schwengels, 1989). With regard to both the slow and rapid change world scenarios (SCW and RCW), there is a marked difference in projected global warming, if all countries participate in stabilizing policies (SCWP and RCWP curves), if only the LDC abstain from participation, or finally, if no countries participate. The stabilising policies are assumed to lead to substantial changes in technologies and fuels.

To summarize, the LDC's first priority has to be the generation of more energy. If clean energy is not affordable to them, they may continue to accept the environmental impacts related to the generation of energy as an added price to be paid for development. What the LDC cannot accept - if that should be the final result of the AIC energy and environment policies for the world - is to be permanently accorded the role of the "epsilons" of Huxley's "Brave New World".

THE PROFESSIONAL INVOLVEMENT

The question of what can be done to limit and reduce the environmental impacts related to energy brings into play almost all professional activities. There are scientific and technological aspects, as well as social, political and economic ones. Even religion and cultural traditions play a role.

Risking oversimplification, one could talk about three main lines of action. One is to improve the efficiency of our current consumption of energy to moderate the demand. Another is to improve the control of the emissions. A third is to transform the entire energy system toward a more rational scheme, by using all potential sources of energy in ways best adapted to the end use in question and to ensure environmental protection. Modifications in the organization of society and in the behaviour of the individual and its perception of "essential needs" may also play a significant role.

A particular study (Davis, 1990) claims, that average efficiencies in end use technologies could be improved by 10 to 25 per cent through the utilization of today's best available practices, as well as gaining an equivalent amount within 20 years as "best practices" improve. For example, advanced technologies based on materials science, engineering, microelectronics and computerization, can produce cars having up to three times the present efficiency level in terms of kms per litre, by the year 2010.

Advances in clean coal technologies - now in progress, e.g. fluidized bed combustion - are expected to improve the efficiency of coal-fired power plants by 10 to 15 per cent, with the advantage that for each 5 per cent efficiency improvement, CO_2-emissions are reduced by approximately 15 per cent. Farther ahead, thermodynamic efficiencies of 45 to 60 per cent may be attained through fuel cells and magnetohydro-dynamic technologies (Kane and South, 1989). Materials science is expected to bear a great effect in the transformation of the energy system (Gray et al., 1990). This does not only apply to the design of new materials for specific energy efficient purposes, but also to the search for ways to recycle economically energy intensive materials (e.g. aluminium, lead, polystyrene, polyethylene, etc.).

The development of energy efficient technologies, as well as of those addressed directly to the control or suppression of unwanted emissions, such as catalytic con-verters for cars or scrubbers for fossil fuel power plants, etc., require the participation of scientists and engineers in almost every field. This also applies to all research which may lead to fundamental transformations of the energy systems, such as nuclear fusion or superconductivity. The preparation of environmental policy instruments, like those shown in Table 3 - or new ones - requires the articipation of legal experts as well as of political and social scientists. Societal behaviour and population growth should be the concern of every individual, not only of the professionally trained.

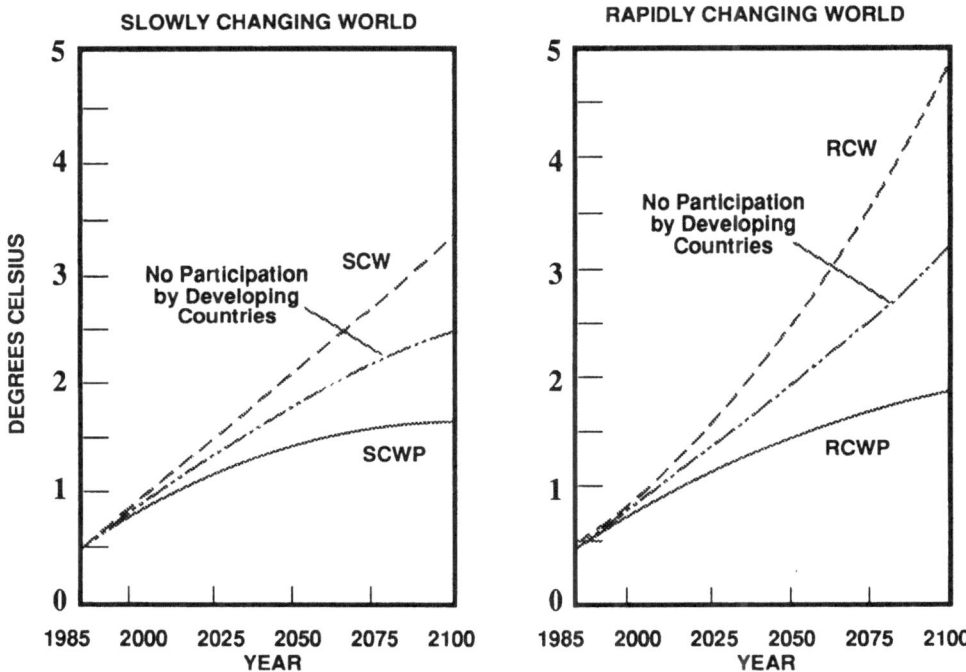

Figure 3. Increase in Realized Warming when Developing Countries do not participate. Degrees celcius: based on 3.0 degree sensitivity. (Source: Schwengels, 1989).

It is essential to not only improve existing technologies and generate new ones, but also to assure their world-wide implementation. This requires new approaches to technology transfer problems, inclusive of training the recipients and granting of appropriate financial assistance. This aspect is crucial to the LDC, but may eventually become so for the AIC as well. The multidisciplinary groups could make an important contribution by identifying, and then clarifying, for the decision makers the advantageous trade-offs between short-term profits and longterm benefits.

Slow development of the LDC may imply less pressure on the world's commercial energy resources and less expenditures for the AIC in seeking new ones, but the political, social and economic consequences may create instabilities which will be far more costly to the AIC. On the other hand, an accelerated development of the LDC certainly demands a considerable increase in their consumption of energy. The required investments necessarily precede by several years any sign of improvement in the standard of living, and, to a large extent, they need to be supplied by the AIC. But eventually, raising the standard of living in the LDC will not only reduce the ecologically damaging use of non-commercial fuels, but will increase the markets for goods and services. Furthermore, as can be clearly seen from the AIC experience, it will slow down the population growth. This is the proper road to a sustainable growth, rather than control measures that infringe upon human dignity and personal freedom.

The funding aspect is essential to any progress in environmental control. One proposal (Grubb, 1989) consists in the assignation of equal emission rights per capita. Based on this, at present the AIC largely exceed their quota, while the LDC still have an allowance. "Emission rights" could therefore be traded. Fixing a carbon tax of U.S. $250 per ton (estimated to eventually render competitive synthetic fuels and alternative non-carbon emission fuels, like hydrogen, by electrolysis from solar or nuclear power), "the selling of 100 million tons of carbon emission rights would represent a financial transfer of U.S. $25,000 million annually" (Manne and Richels, 1990). The ethical aspect of such a proposal is questionable, if interpreted as a right to pollute when willing to pay.

Another example of a concrete proposal, addressed to the problem of stabilizing the atmosphere at the present level in terms of greenhouse gases, is the one outlined below (Goldemberg, 1990). Based on a study prepared for the Ministerial Conference on Atmospheric Pollution and Climatic Change, held in Noordwikj in 1989, a requirement of U.S. $50,000 million per year is estimated. To raise that amount, a worldwide CARBON TAX is proposed, related to energy consumption at a rate of U.S. $6 per Toe. Therefore, most of the burden would fall on the AIC - and about 50 per cent of this on the OECD - but the LDC would also contribute. About two fifths would go to the AIC, in order to incite additional energy efficiency measures and to support a gradual withdrawal of CFCs. The rest would be a grant fund (not loans - there is enough debt already) for the LDC, to be applied essentially for: 1) leapfrogging development policy; 2) transferring to the LDC technologies for an environment safeguarding development; 3) reforestation.

Table 3
Environmental policy instruments (Source: Davis, 1990).

INSTRUMENT	APPLICATIONS (examples)
Economic instruments	
Taxes (Polluter-pays-Principle)	☐ Greenhouse Gases ☐ Acid Deposition
Subsidy/Tax relief	☐ Energy conserving Investments ☐ Energy research (private) ☐ Unleaded fuels: catalytic converters
Direct public investment	☐ Afforestation programmes ☐ Energy research (private)

Grant aid	☐ LDC afforestation, soil conservation,
	   agrarian reform and land settlement
	   programmes
	☐ Regional programmes in industrial
	   countries

Law	
Public regulation: Law, legal liability	☐ CFCs ☐ Toxic wastes ☐ Community property rights ☐ Unleaded fuels: catalytic converters

Negotiation	☐ Localised issues: Emissions, wastes,
	   compliance with local land use and
	   building codes

Values	
Self-regulation	☐ "Green" consumers and producers
Positive Side-effects of Policy	
Economic policies with environmental benefits	☐ Urban congestion charges ☐ Forestry Policy (LCDs) ☐ Agrarian reform (LCDs)

The above-mentioned points one and two are closely related. Instead of trans-ferring obsolete technologies to the LDC, especially in the still needed energy intensive sectors like steel or power plants - but also in the consumer goods industries - the grant would partially take care of the added capital expenditure for energy efficient designs and pollution control equipment being developed mostly in the AIC. An example of the importance for the LDC to leapfrog obsolete or inefficient technologies can be found in the field of refrigeration. The availability of cold storage, both at industrial and domestic levels, can bear a considerable world-wide influence on food availability by diminishing losses as well as by preserving hygiene and reducing health expenditures arising from food intoxications. The AIC possess at present 2/3 of the industrial cold storage capacity and the availability of domestic refrigerators can be considered universal. The LDC, among which many are located in the tropics, are deficiently supplied in both of these aspects. Thus, it makes both economic and social sense to promote the availability of cold storage technologies and appliances. These will create, however, a demand for electricity. In view of the several million refrigera-tors that should be incorporated annually in the LDC, an appliance consuming 400 kWh/year instead of 800 kWh/year can make a considerable difference in the required electric capacity to be installed (Goldemberg, 1990).

Finally, we should emphasize another very important aspect where the AIC can help the LDC to advance more rapidly in their development. There is an obvious need of education and training at all levels, because the LDC are at many different stages of development. However, the possibility of leapfrogging is closely related to the following: an active participation of the LDC in the process of research and develop-ment of future energy technologies.

Indeed, a crucial difference between the developing countries - even among those who already possess a large industrial base - and the advanced countries is that the latter have the know-how to create additional know-how. Unless the LDC can begin to acquire this basic skill, no true development will ever be achieved. This is the main challenge for development.

CONCLUSION

Energy is necessary to reach an acceptable standard of living. The consumption of energy gives rise to severe environmental impacts. Deterioration of the environment constitutes a threat to the standard of living and, in the long run, to life itself.

The interweaving of energy with world economics, geopolitics, social develop-ment and environment constitutes a complex structure in a multidimensional space. Any cross section, however carefully attempted, is liable to distort or even neglect important features, to the extent of sometimes leaving out the parameters most signifi-cant to the aspect one is attempting to focus on. Short-term and long-term views often give contradictory perceptions of what the problems are, if these can be clearly recog-nized at all.

The clarification of issues and the creation of solutions involve practically all fields of professional expertise. However, the various efforts should keep a world-wide perspective in view. No single country or society is responsible for the energy related environmental impacts, nor will it be capable of keeping itself isolated from the rest of the world and solve its own problems. Only through cooperation a confrontation between energy needs and environment can be avoided and a solution perhaps - or at least a satisfactory compromise - can be achieved.

REFERENCES

Davis, G.R. (1990), "Responding to the Challenge of Global Warming: The Role of Energy Efficient Technologies". In Proceedings of an Experts' Seminar on Energy Technologies for Reducing Emissions of Greenhouse Gases,
Vol. 1: 71-85. OECD/IEA.

Dixon, J.A., Talbot L.M., and Lemoigne G.J-M., (1989), "Dams and the Environment". World Bank Technical Report No. 110, Washington.

DOE/EIA (1990) "International Energy Outlook 1990",
Washington: DOE/EIA - 0484(90).

Goldemberg, J. (1990), "Energy and Environmental Policies in Developed and Developing Countries". In Energy and the Environment in the 21st Century, Preprint.
Vol. I: 69-100. Massachusetts Institute of Technology, Energy Laboratory, Cambridge, MA 02139.

Gray, P.E., Tester, J.W. and Wood, D.O. (1990), "Energy Technology: Problems and Solutions". In Energy and the Environment in the 21st Century, Preprint Volume I. Massachusetts Institute of Technology, Energy Laboratory, Cambridge, MA 02139.

Grubb, M. (1989), "The greenhouse effect: negotiating targets", The Royal Institute of International Affairs, London.

Hartley, D.L. and Schueler, D.G. (1990), "Perspectives on Solar Energy and the Environment". In Energy and the Environment in the 21st Century, Preprint. Vol. I, Massachusetts Institute of Technology, Energy Laboratory, Cambridge, MA 02139.

Kane R. and South, D.W. (1989), "Potential fossil-energy-related technology options to reduce greenhouse gas emissions". In Proceedings of an Experts' Seminar on Energy Technologies for Reducing Emissions of Greenhouse Gases,
Vol. 1: 35-58. OECD/IEA.

Manne A.S. and Richels, R.G. (1990), "Global CO_2-Emission Reductions. The impacts of rising energy costs". In Energy and the Environment in the 21st Century, Preprint Volume I. Massachusetts Institute of Technology, Energy Laboratory, Cambridge, MA 02139.

Marques de Souza, J. A. and Bennett, L. L. (1989), "Nuclear power for environmental protection". In World Energy Conference. 14th Congress, Division 2: Energy and the Environment (WEC, ed.), paper 2.2.15.

OECD (1983a) "Environmental Effects of Energy Systems", (OECD, ed.), Paris.

OECD (1983b), "Coal and Environmental Protection: Costs and Costing Methods", (OECD, ed.), Paris.

OECD (1985) "Environmental Effects of Electricity Generation", (OECD, ed.), Paris.

OECD (1986) "Environmental Effects of Automotive Transport: The OECD COMPASS project", (OECD, ed.), Paris.

OECD (1987), "Coal and Environmental Policies and Institutions", (OECD, ed.), Paris.

San Martin, R.L. (1989), "Environmental Emissions from Energy Technology Systems: the Total Fuel Cycle". In Proceedings of an Experts' Seminar on Energy Technologies for Reducing Emissions of Greenhouse Gases, Vol. 1., OECD/IEA.

Schipper, L. (1990), "Energy Saving in the U.S. and Other Wealthy Countries: Can the momentum be maintained?" In Energy and the Environment in the 21st Century, Preprint Volume I, pp. C6-C21. Massachusetts Institute of Technology, Energy Laboratory, Cambridge, MA 02139.

Schmidt F., (1989) "Competitive Distortions due to Varying National Environmental Conservation Standards". In World Energy Conference. 14th Congress, Division 2: Energy and the Environment (WEC, ed.), paper 2.1.5.

Schwengels, P. (1989), "Energy Technology Options for Climate Warming: preliminary results". In Proceedings of an Experts' Seminar on Energy Technologies for Reducing Emissions of Greenhouse Gases, Vol. 1: 3-20. OECD/IEA.

WCED (1987), "Our Common Future", Report by the World Commission on Environment and Development ("Brundtland Report"). Oxford University Press.

WEC (1988), "Environmental Effects Arising from Electricity Supply and Utilization and the Resulting Costs to the Utility". World Energy Conference Report 1988, p. 11.

WEC (1989), Proceedings of The World Energy Conference. 14th Congress (WEC, ed.).

WEC (1990), "Solar Power", World Energy Council Report 1989, p.A1-6.

Wood, T.S. and Baldwin S. (1985), "Fuelwood and Charcoal Use in Developing Countries". Annual Review of Energy 10: 407-429.

World Bank (1989) "World Development Report 1989". Oxford University Press.

7.

METEOROLOGY

AKSEL C. WIIN-NIELSEN
Geophysical Institute
University of Copenhagen
Haraldsgade 6, DK-2200 Copenhagen N,
Denmark

ABSTRACT

The development of steadily larger concerns of an environmental nature in the atmospheric sciences is described. The environmental edge in meteorology has become more and more important due to developments which have taken place over the last half century or so. A great variety of environmental problems have appeared in a number of areas, and it will be impossible to cover them in a single short paper. The present paper will, therefore, concentrate on problems related to climate and climate change, and within this general area, on the greenhouse effects.

A comparison will be made between the results of model studies and studies of the available climatological data with the major conclusion that, so far it has not been possible to create a direct link between the steadily increasing concentration of the greenhouse gases, particularly carbon dioxide, and the observed changes of temperature at the surface of the Earth. On the other hand, data from the upper atmosphere seems to indicate that the greenhouse gases may have had an effect on the temperatures above the major cloud layers and especially in the lower stratosphere.

Key Words: *Climate Change; Climate Change, Model Predictions of; Climate Change, Observational Studies of; Climate Models; Greenhouse Effect, general; Greenhouse Effects; Greenhouse Effect, Special.*

INTRODUCTION

Once upon a time, not so long ago, environmental issues were few and far between in the field of meteorology. Consulting the Compendium (1951), published by the American Meteorological Society and giving a picture of the important issues of meteorology around the middle of the century, one finds only one chapter containing a single section on Air Pollution. This paper was concerned mainly with the emissions from a single source or multiple sources quite close to each other. The important concern was the measurements of air pollution in a single community and the description of the problem of pollution using diffusion equations. We may, thus, conclude that the local issue of pollution was the major concern at that time.

Since then a large number of environmental issues have entered the broad field of atmospheric sciences. While only a small fraction of the atmospheric science community was concerned with pollution problems around the middle of the century, today we find that very few atmospheric scientists can avoid becoming involved in one issue or another. The reasons partly result from the fact that the issues require the use of the most advanced tools available in the field of atmospheric sciences; and partly from the fact that the potential impact on society has increased considerably. Consequently, the appearance of these various environmental issues has caused new areas of speciality to be created. Typically we find that a new area starts as interdisciplinary research but gradually, in the course of a few decades, these boundary topics are converted to new subdisciplines anchored within the atmospheric sciences. A typical example would be the field of atmospheric chemistry which started as a cooperative effort between meteorologists and agronomers who were interested mainly in the chemistry of precipitation. Today the field of Atmospheric Chemistry is taught at major universities and it is becoming increasingly important in the treatment of major environmental issues.

As indicated above the environmental issues are now so numerous that it will be impossible to treat them in detail in a paper of very restricted length. In addition, the author is far from qualified to handle the details of many of them. In the following we shall, therefore, be quite selective and shall prefer to concentrate on a single issue, i.e. the potential effect of the increase of some of the greenhouse gases such as CO_2, but other important environmental topics should at least be mentioned in these introductory remarks.

Pollution problems are no longer treated as local issues. It has been realized that polluted air may be transported over long distances and may be deposited far from the source. This general problem, which enters in quite a few specific areas, has led to the development of methods to compute the three dimensional trajectories of air parcels. The problem had already appeared in the 1950's in connection with the transport of fertilizers entering the atmosphere by wind erosion and deposited elsewhere. A preliminary solution (Djuric and Wiin-Nielsen, 1956) took advantage of the fact that the vertical velocity in large-scale motions is typically two orders of magnitude smaller than the horizontal velocity, and a trajectory based on the observed horizontal winds should, thus for relatively small time periods (12 to 24 hours), be a reasonable approximation. With the development of global analyses in the 1970's and good methods to determine the vertical velocity diagnostically from three dimensional atmospheric models, the computation of both diagnostic and prognostic trajectories can now be performed at the major meteorological centres of the world. The use of diagnostic trajectories, computed backward in time, has been very important in determining the source regions for pollutants of various kinds.

A major environmental issue appeared as a result of the energy crisis in the 1970's. Due to the shortage of oil alternative energy sources were naturally investigated. One possibility was to use wind energy to produce electrical power. Over the past two decades or so, the use of wind energy created by windmills has become a reasonably popular alternative source of energy in those parts of the world where the wind conditions are favourable. Meteorologists have assisted in the solutions of these problems by providing wind atlases (Troen and Petersen, 1989) designed specifically for the windmill operators. These atlases are designed in order to provide guidelines and relevant statistics which can be used in the computations of wind energy resources in a certain region.

Another relatively recent environmental issue is acid rain. It is a by-product of the industrial processes in which sulphur derivatives, apparently mostly sulphur dioxide, SO_2, and various NO_x enter the precipitation, since they are dissolved in the water droplets. The resulting acid rain may have a profound effect on the vegetation and on the acidity of rivers and lakes. The issue has been severe in certain parts of the world, notably Central Scandinavia and Southern Canada. Meteorological trajectories have been helpful in determining the source regions in both particular cases and in a climatological sense.

The ozone issue is the most recent environmental issue which has appeared. As with the previous issue, it has strong overtones of atmospheric chemistry because its theory is that man-made chemicals, especially freon, should destroy the ozone which is mostly located in the lower part of the stratosphere. Additionally, as was the case in the previous issue, it is not possible to come to a real understanding of the problem without a combined effort from the atmospheric chemists and the atmospheric dynamicists. Unfortunately, the space available in this paper does not permit going into details.

In the remaining part of this paper we shall, as mentioned above, deal with a single environmental problem which is considered to be the most important issue of all those which have had a profound influence on the atmospheric science community in the last decade or so, i.e. climate and climate change due to anthropogenic effects.

In this regard it is important to distinguish between the general and the special greenhouse effects. The general greenhouse effect defines the influence on the climate due to the fact that the atmosphere contains such gases as CO_2, while the special greenhouse effect deals with the changes in the climate due to changes in the concentration of the various greenhouse gases. While the rest of the paper will deal mainly with the possible effect of these changes in the concentration of CO_2, it should be stressed that, in the space allowed for this paper, one cannot describe the other gases which may have a pronounced effect on the radiation budget of the atmosphere.

THE CLIMATE SYSTEM

In order to be able to appreciate the following considerations, it will be necessary to remind the reader of a few facts.

The climate, at any given time, is defined officially (World Meteorological Organization) as "the manifestation of weather over a particular area and for a particular period of time". The climatological normals are averages calculated for consecutive 30-year periods as for example the 1st of January 1931 to the 31st of December 1960. The first determination of the Earth's climate, based on real observations, covers this period. We note in passing that a new climate period is about to be finished (1961--1990) and we look forward to an intercomparison of the two periods. The climatic state of the atmosphere, in past times, is a reconstruction based on indirect information from many different sources. For example, the climate over the last 1,000 years is partly determined by historical sources and partly by more objective measurements (i.e. ice cores and cores taken from the bottom of the sea). These core measurements can also be used to reconstruct the climate for many thousands of years. It is by such methods that we have gathered information on glacial and interglacial periods. These measurements tell us that the global mean temperature has not changed by more than 5-8 °C between a glacial and an interglacial time.

While we are dealing with the climate of the atmosphere, and especially the climate of the lowest few meters close to the surface of the Earth, we should realize that the climate is under a strong influence from the interactions between the atmosphere and a number of other systems such as the oceans, the continents, the ice masses on the continents and in the oceans, and the living material on the Earth, normally called the biomass. A real understanding of the climate is not possible without considering all these interactions which, incidentally, are not too well understood, at least not all of them.

CLIMATE MODELS

It has become customary to build models of the climate. These models may vary from very simple models for educational purposes to extremely complicated models containing the total available information of the climate system. The complicated models have their origin in weather prediction models which were developed for forecasting purposes beginning in the 1950's. The first purpose of these models has always been to simulate the present general circulation of the Earth's atmosphere but, gradually, they have become known as climate models. The first attempts to simulate the general circulation of the atmosphere were made by Phillips (1956) and Smagorinsky (1963).

Ideally a climate model should describe all the significant physical processes which go on within and between the components of the climate system, i.e.

- the atmosphere
- the oceans
- the surface of the continents
- the ice masses on the continents and in the seas
- the biomass.

Even after 3 decades of work on the climate models, however, it must be stressed that we do not yet possess models which include all these processes and components in a satisfactory way. Originally the models were purely atmospheric. The next step was to incorporate part of the ocean-related processes in a crude fashion. It was realized that the interactions between the oceans and the atmosphere with respect to heat, moisture and momentum were important in driving the atmosphere which, to a large extent, is driven from below, although the primary energy source naturally is the energy received from the Sun. It was also clear that the time scale of the oceans is much larger than the time scale of the atmosphere, or, said in another way, the motions in the oceans are generally two orders of magnitude slower than those in the atmosphere. Consequently, it was assumed that the oceans were without motion, and that the influences of the ocean on the atmosphere could be accounted for by specifying the surface temperature of the oceans (from measurements), because this parameter is sufficient together with atmospheric parameters to calculate the important interactions. While these assumptions are justified for prediction purposes for one or perhaps two weeks, it should also be clear that, for much larger time scales, we cannot disregard the current in the oceans which is responsible for the changes in the sea-surface temperatures due to advection. A real climate model must, thus, include a coupled ocean-atmosphere system. It is somewhat easier to include the continents and their ice masses, since their motions can be disregarded for all but the very largest time scales. The problems of incorporating the sea ice is again difficult because the sea ice, when in motion, is determined by oceans currents and by the atmospheric winds. In addition, sea ice can be generated in the northern latitudes and can melt in middle latitudes.

Time is not available to describe the particular details of the models. The advanced models, however, are based on a set of non-linear differential equations with one equation for each variable. These equations describe the physical processes incorporated in the model.

In the description of the physical processes in the climate system we are faced with the dilemma that almost all of them take place on a micro-scale. The models have to be described in terms of the variables present in the model, however. They describe a macro-scale much larger than the scale on which the real processes take place, but as we are unable to describe what happens with each molecule or atom, the description of the various processes in terms of the macrovariables is called a "parameterization", because we express the processes in the available parameters. This situation is well known in the branches of physics, the classical description of the thermodynamics or the treatment of turbulence are examples. In any case, the procedure is far from satisfying because one is always left with the impression that a given parameterization could be improved. Due to this situation, a great deal of time is used in the modelling groups to attempt to improved the descriptions but we are aware that the answers are forever going to be inexact.

The processes may be listed as follows:

-    the radiation budget including the clouds, the moisture distribution, the trace gases and the nature of the underlying surface of the Earth;

-    the convection prescription which is one of the many processes where the real physics takes place on a micro-scale, but has to be described in terms of the macro-scale variables. The general name is "parameterization" for such a procedure;

-    the large-scale precipitation prescription for the precipitation connected with fronts in atmospheric systems;

-    the prescriptions for the exchanges of sensible heat, evaporation and momentum between the underlying surface (land, water, ice) and the atmosphere;

-    the dynamical influence of topography (mountains) on the atmospheric flow;

-    the prescriptions for turbulence in the atmospheric boundary layer and in the "free" atmosphere;

-    a detailed description of the land surface with respect to vegetation including the albedo for each location;

-    a prescription for the treatment of the soil moisture.

THE USE OF THE MODELS

The models have been designed and used mainly for the simulation of the present climate of the atmosphere. We may say briefly that they are rather good in simulating the zonally averaged state of the atmosphere and less good in the simulation of the deviations from the zonal state, i.e. the wave regime in the atmosphere. The accuracy of the simulation depends naturally on the ability to use accurate prescriptions, but is also a strong function of the resolution in the model. The resolution, vertically and horizontally, had to be small in the early days due to lack of computer storage and speed, but today it is normal for prediction models to have grid sizes of less than 100 km in each "horizontal" layer and a number of layers exceeding 20 to describe the vertical structure of all the variables. Further improvements in computational speed are necessary in order to permit the same resolution in climate models because they must simulate much longer time periods. A typical climate model may have a grid size of 250 km and about 10 layers.

A large number of sensitivity experiments have also been carried out. Typically, we are interested in knowing how much the final result is changed for a given change in one of the external conditions or in the testing of a new procedure or a prescription. Of great interest here is the use of the models in simulating a climate change or a climate variation. The best example to be selected is the special greenhouse effect, and we shall, in a later section, give the results of such sensitivity experiments.

THE GREENHOUSE EFFECT

We normally distinguish between the general greenhouse effect, which is the effect of the atmosphere on the surface temperature of the Earth, and a special greenhouse effect, which is the changes in the temperatures and other parameters in the atmosphere due to changes in the concentrations of the components of the atmosphere.

It is quite easy to give a simple model illustrating the general greenhouse effect. The following model serves one purpose only: to illustrate the radiative processes at work. For this purpose we look at the Earth-atmosphere system as a single layer having a single global mean temperature T_a. The mean temperature at the surface of the Earth is defined as T_e. The radiation received from the Sun per unit area at the outer rim of the atmosphere will be called $Q = 344$ Wm^{-2}. We define the planetary albedo, α, as the fraction of Q which is reflected back to space without participating in the atmospheric heat budget. This means that $(1-\alpha)*Q$ is left for use by the atmosphere. We shall assume that this quantity goes through the atmosphere and provides an input to the radiation balance at the surface. The surface of the Earth will radiate σT_e^4 upward. Here $\sigma = 5.67 \times 10^{-8}$ is the Stefan-Boltzman constant. As a simple assumption we assume that all the long wave radiation from the Earth is absorbed in the atmosphere which itself radiates an amount σT_a^4 both upward and downward. With the above assumptions we may write the following two equations for the surface and for the atmosphere, respectively:

$$(1-\alpha)Q + \sigma T_a^4 = \sigma T^4 \qquad (1)$$

$$\sigma T_e^4 = 2\sigma T_a^4 \qquad (2)$$

With the planetary albedo $\alpha = 0.3$, it is possible to solve the equations with the result that $T_e = 303$ °K and $T_a = 255$ °K. If there was no atmosphere, and if the planetary albedo was unchanged, then eq. (2) would disappear and so would the second term on the left hand side of eq. (1). In that case the surface temperature, which becomes $T_e = 255$ °K, thus would be the temperature of the Earth if the atmosphere was not present. In other words, due to the existence of the atmosphere, the simple model (which provides a rough estimate only) gives a surface temperature, which is about 50 °K higher than it otherwise would be.

We may use the simple model to illustrate sensitivity studies. Only two parameters enter the model: the solar input and the planetary albedo. Many older theories have proposed that changes in the solar output are responsible for climatic changes. From our model, it is straightforward to calculate that if the global mean temperature should change by one degree °K, it will be necessary for the quantity Q to change by an amount of 4.5 Wm^{-2} corresponding to a percentage change of 1.3 per cent. Since we are working with steady states we require such a change in the mean value of Q, not in the fluctuations around the mean value. We add also that changes in this amount in the mean value of Q have not been observed.

It is known that the change in the global mean value of the temperature between a glacial and an interglacial period is not larger than 5-8 °K. For an albedo value of 0.30, we found a surface temperature of 303 °K. Repeating the calculation with an albedo value of 0.35, corresponding to a larger ice cover, we find a mean temperature of 298 °K, while a decrease of the albedo to 0.25 results in a temperature of 309 °K. There is, thus, reasonably good agreement between the extent of the ice cover and the resulting global mean temperatures.

Somewhat more complicated models are necessary to illustrate the special greenhouse effect which space will not permit a detailed account of. These models have been summarized by Schlesinger (1988).

THE SPECIAL GREENHOUSE EFFECT

The great concern in the last couple of decades in the studies of possible climate change has been the possibility that the temperature of the troposphere could increase as a result of the steadily increasing concentration of CO_2 in the atmosphere. The increase is due to the burning of coal, oil and gas for energy purposes. The concentration was about 270 ppm in the preindustrial period, around 1800-1850. Estimates by Rotty (1983) show that the emission has been about 4.6 per cent per year in the period 1860-1973. Since 1973, the increase in the emissions has continued at a diminished rate of about 2.3 per cent per year. It should be stressed that only a fraction of these emissions remain in the atmosphere. With respect to the increase in the concentration of CO_2 in the atmosphere, it has been estimated that the increase has been 0.11 per cent per year for the period 1860-1973 and 0.46 per cent per year for the time interval 1973-1990. Estimates indicate that a concentration twice as large as the preindustrial level will be reached sometime in the 21st century. The question is: How much and in which way will the climate change due the special greenhouse effect?

The first study of CO_2-induced temperature change was carried out by Manabe and Wetherald (1967). They used a model containing variations in the vertical direction only and performed the calculation with three different concentrations of CO_2: 150, 300 and 600 ppm. Increases in the concentrations were accompanied by increases in the temperature of the troposphere and decreases in the stratosphere. While the temperature increase was only a couple of degrees at the surface of the Earth, considerably larger decreases were found in the stratosphere.

Since then numerous simulations of the special greenhouse effect have been made using three-dimensional models. The typical experiment consists of performing one time-integration of the model equations with the present concentration of carbon dioxide followed by a second integration having, say, twice the concentration. Since the change in the concentration of CO_2 is the only difference between the two experiments, it is assumed that the difference in the end results can be ascribed to the doubling of the carbon dioxide concentration. Later experiments have used a steadily increasing concentration simulating the real increase in the atmosphere. The results of such experiments have been widely used to make statements on the atmospheric impact of the steadily increasing carbon dioxide content in the atmosphere. However, as seen from Fig. 1, it is also evident that different models give different results. The estimates from the various models vary from 1.9 °K to 5.2 °K or almost by a factor of 3. We also notice that the models with a relatively small mean surface temperature give the largest changes. While an explanation might be that the smaller the surface temperature, the larger the extent of the ice and, thus, the larger the positive feedback, it must be stressed that such an explanation cannot explain the very large differences in the estimated increases. The explanation is, thus, found in the differences in the parameterizations employed in the various models.

The models also give information about other meteorological parameters. An example is given in Fig. 2 where the percentage increase in the precipitation is plotted against the estimated increase in the surface temperature. Here we find large differences between the models with the lowest change being 4 per cent and the largest about 15 per cent. Papers on model results are numerous and they contain a large number of results and figures. According to Mitchell (1989), who has produced several reviews and comparisons, we may summarize the major results in the following way:

- the stratosphere will cool
- the troposphere will warm
- the global water cycle will be enhanced.

While these results are common to all models and therefore judged to be certain, he is less certain about the following statements, but consider them to be "probable":

- the warming will be largest in high latitudes in winter
- the annual mean warming will be smallest in the tropics with little seasonal variation
- precipitation increases in middle and high latitudes in winter
- seasonal sea-ice will melt earlier and will form later.

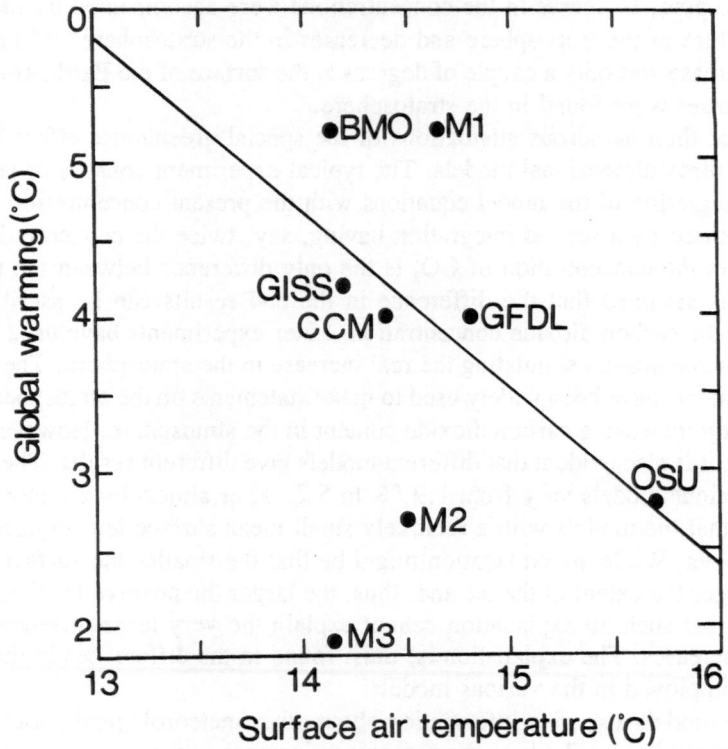

Figure. 1. The dots indicate the computed global warming in °C for various model experiments as a function of the global mean surface temperature also in °C. The abbreviations have the following meaning: BMO: British Meteorological Office, CCM: Climate Community Model (National Center for Atmospheric Research), GISS: Goddard Institute for Space Sciences, GFDL: Geophysical Fluid Dynamics Laboratory (Princeton, New Jersey), M1, M2 and M3: Mitchell's experiments No. 1, 2 and 3, OSU: Oregon State University. (After Mitchell, 1989).

As one can see from the very brief review given above, the state of affairs in the simulations of the special greenhouse effect is far from satisfactory. As long as estimates using different models vary as much as we have seen, it is difficult to have faith in the results. The model designers agree on this point, and all are working to produce better models by incorporating better parameterizations. Of major concern is the estimates of the role of clouds in the special greenhouse effect. While it is true that the direct effect of the increased concentration is accounted for in the radiation budget, the models have had difficulties in describing the secondary feedback mechanisms which result from the computed warming. Such a mechanism could be an increase in the cloud amounts, which in turn could give a decrease in the layer below the clouds. The large differences in the simulated tropical temperatures point, thus, to the importance of cloud and convection prescriptions. And, above all, there is a long way to go before the biomass and all the processes connected with it are represented in an adequate way in the models.

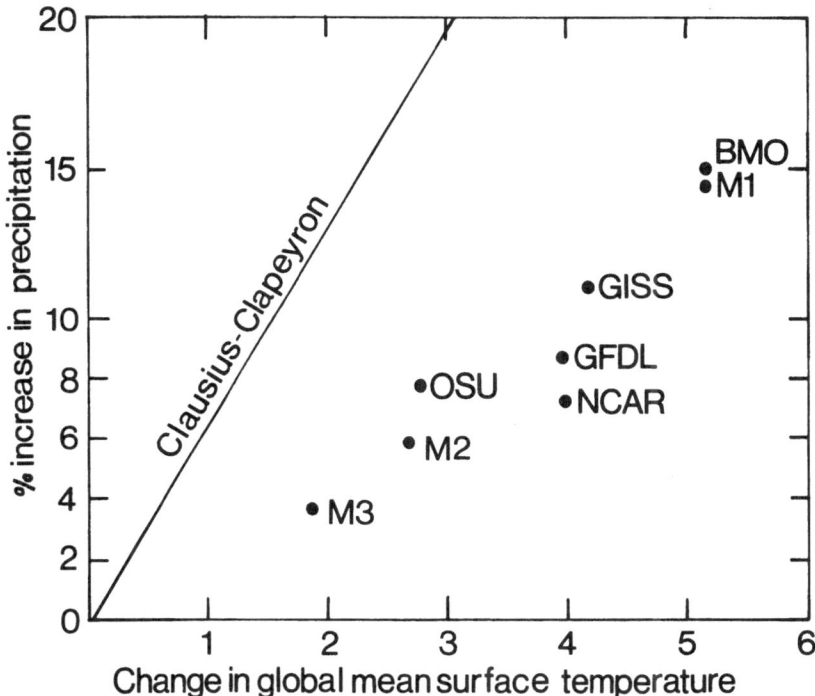

Figure 2. The percentage increase in precipitation as a function of the change in the global mean surface temperature for the various models. Abbreviations as in Fig. 1. (After Mitchell, 1989). The curve marked Clausius-Clapeyron shows the precipitation increase, which would be experienced, if the water vapour made possible by the increase of temperature, would fall out as precipitation. (After Mitchell, 1989).

DATA STUDIES

While model studies have been going on for more than two decades, it is only within the last few years that climatologists have taken an interest in whether or not the results from the simulations can be verified from the time series of climatological data. The reason for this delay is understandable simply because climatological data were traditionally arranged station by station. It therefore took some time before hemispheric and global data sets of the surface climatological data were arranged (Jones et al., 1986). This group prepared data sets for both the northern and the southern hemisphere. From such data sets, which cover about 100 years before the present, they concluded that a temperature increase at the surface of the Earth amounted to about 0.5 °K for 100 years. The problem is now whether or not such a temperature change can be ascribed to the special greenhouse effect?

In addition to the greenhouse effect we have to look at other reasons for a change in the climatological global averages. Two essential processes come to mind. The first is the influence of volcanic dust in the atmosphere. This problem has been investigated by Bradley (1988). He comes to the conclusion that the content of volcanic material in the atmosphere after a volcanic eruption will result in a cooling on the global scale of 1 to 2 °K. He also states, however, that the effect seems to disappear after about 1.5 to 2 years when by far the greater part of the volcanic material has been removed from the atmosphere where the residence time is about a week for the troposphere and 1-2 years for the stratosphere. The volcanic effect is, therefore, a cooling but, under normal volcanic activity, it will hardly influence the long term trend. The other factor is the so-called urban effect which stems from the fact that few climatological stations have remained in the same place, and even if they have, it is most likely that the immediate surroundings have changed significantly during the years. A typical example is a station which was established more than a century ago well outside a city. In the meantime the city may have grown to such an extent that the station now is well within the densely populated area in an environment dominated by tall buildings, concrete surfaces etc. If so, the local temperature, precipitation and wind will be quite different from what they would have been if the surroundings of the station had been unchanged. Wood (1988) was very critical of the way in which Jones et al. had tried to correct for the urban effect. His reason being that they had tried to discover the urban effect at a given station by comparing that station with one or several stations in the neighbourhood but, as Wood points out, both stations will stay in the sample if they are both under the influence of the urban effect. The conclusion is, therefore, that the total urban effect has not been removed from the data sets and that the estimated rise of the temperature by 0.5 °K per century is too large.

In view of this discussion and considering, furthermore, that the climatological data over the oceans are quite uncertain, Hanson et al. (1989) decided to investigate the area-averaged data for the United States excluding Alaska and Hawaii. The time series of the data were investigated to see if a significant change in the long term mean could be discovered. The procedure was to use the "Spearman Rank Test" and "two-phase regressions". The result was that no significant change in the long term mean could be found. A later investigation by Karl and Jones (1989) came to the conclusion that the urban effect over the continental USA amounted to 0.1 to 0.4 °K. Correcting the earlier data they arrived at the conclusion that the part of the observed temperature rise which can be ascribed to the greenhouse effect is 0.16 °K.

Climatological surface data are notoriously difficult to use. It has, therefore, occurred to some investigators, including the author (Wiin-Nielsen, 1990), that one could use the operational global analyses made every six hours by a few operational centres in the world. The advantages are a very consistent data set with plenty of vertical resolution and with no urban effects. The disadvantages are that the analyses are, to some limited degree, model dependent and that the time series are too short since it is necessary to limit them to periods within which the same analysis system has been used. For this reason it has been decided to reanalyse the past data with an unchanged system. Nevertheless, the attempt has been made and Fig. 3 and Fig. 4 give the results showing a warming trend at 700 and 500 hPa and a cooling trend at 1000 hPa and 100 hPa. This pilot study will be expanded in the future to cover additional years. A possible interpretation of these preliminary results is that the data in the stratosphere show the expected cooling, but the warming at the surface cannot be observed. We speculate that this in reality is due to an incorrect parameterization of the interplay between the radiation and the clouds. However, a longer data series will have to be produced to make the results more certain.

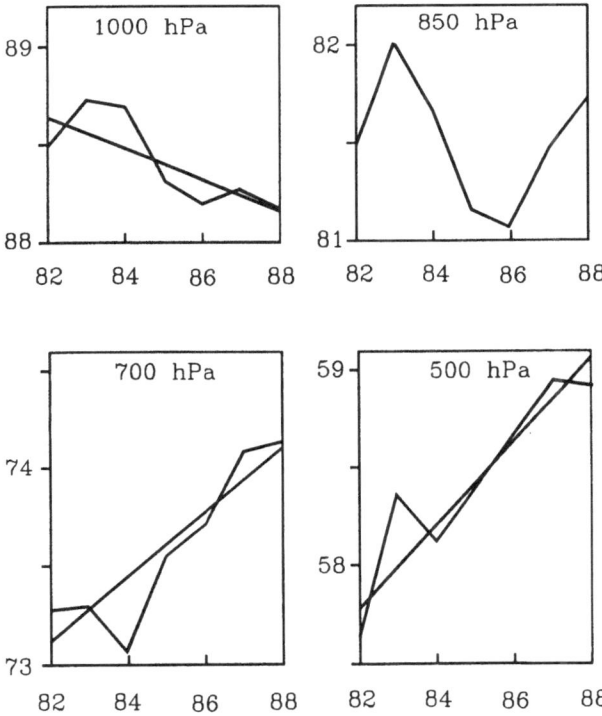

Figure 3. The annual global mean temperatures for the isobaric surfaces 1000, 850, 700, and 500 hPa for the years 1982-1988, incl. The lines in the three figures for 1000, 700 and 500 hPa are linear regression lines.

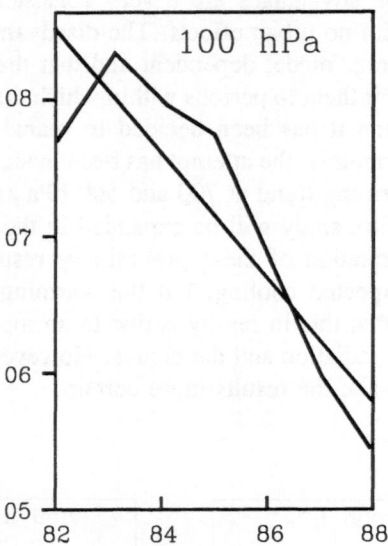

Figure 4. As Fig. 3, but for the 100 hPa surface.

CONCLUDING REMARKS

The special greenhouse effect is a complicated issue. The very fact that the increase of concentration of CO_2 in the atmosphere continues is a threat to the environmental system. The model calculations have tried to calculate the resulting changes in the climate with a variety of results, all of which, however, show an increase in the global mean of the surface temperature (to take a simple example). The calculated changes in this temperature for a doubling of the CO_2 concentrations are between one and several degrees. A major dilemma is that the standard surface data for the last century or so show a temperature increase of at most 0.5 degrees, and this number may even be too large. The application of global analyses produced four times a day by the European Centre for Medium-Range Forecasts is in principle better but is hampered by the fact that the time series are too short at the moment to provide a real climatological data series. Data for the last few years, however, show that the surface data from this source does not demonstrate a statistically significant increase in the global mean surface temperature.

The latter data provide an opportunity to investigate changes in the global mean temperatures for other levels in the atmosphere. From the short record available it was found that the only significant increases occur at heights of 3 to 5 km, while a significant decrease was found at levels in the lower part of the stratosphere. These conclusions are necessarily of a preliminary nature due to the shortness of the records, but should be followed up in the coming years. Were the findings, however, correct, they would indicate that the models do not simulate the physical processes connected with the special greenhouse effect with sufficient accuracy. At the moment some models are being redesigned to include more realistic feed-back processes.

In this paper we have covered only the special greenhouse effect in detail. The many other environmental issues in the atmospheric sciences have been disregarded. The reason is not that they are uninteresting, but only that time and space did not allow their inclusion.

Note Added in Proof:
Since the preparation of the present paper the so called IPCC report has been published (Climate Change, The IPCC Scientific Assessment, Editors J.T. Houghton, G.J. Jenkins and J.J. Ephraums, Cambridge University Press, 1990, 364 pp.). It contains a wealth of material on the topics treated above as well as a treatment of all the other greenhouse gases.

REFERENCES

Bradley, R.S. (1988), The explosive volcanic eruption signal in the Northern Hemisphere continental temperature record. Climate Change, 12: 221-243.

Djuric D. and A. Wiin-Nielsen (1956), Preliminary report on computations of trajectories by means of electronic computers. Berichte des Deutschen Wetterdienstes, No.38: 22-23,

Hanson, K.,G.A. Maul and T.R. Karl (1989), Are atmospheric greenhouse effects apparent in the climatic record of the contiguous U.S. (1895-1987) ? Geophys. Res. Letters, 16, No.1: 49-52.

Hewson, E.W. (1951), Atmospheric Pollution. In Compendium of Meteorology (T.F. Malone, ed.), American Meteorological Society, Boston, Mass., 1317 pp., pp. 1139-1160.

Jones, P.D.,S.C.B. Raper, R.S. Bradley, H.F. Diaz, P.M. Kelly and T.M.L. Wigley (1986), Variations in surface air temperatures, 1851-1984. J. Clim. Appl. Meteor., 25: 161-179.

Karl, T.R. and P.D. Jones (1989), Urban bias in area-averaged surface air temperature trends. Bull. Amer. Meteor. Soc., 70, No.3: 265-270.

Manabe, S. and R.T. Wetherald (1967), Thermal equilibrium of the atmosphere with a given distribution of relative humidity. J. Atmos. Sci., 24: 241-259.

Mitchell, J.F.B. (1989), The equilibrium response to doubling atmospheric CO_2. Report DCTN 82, British Meteorological Office, 20 pp.

Phillips, N.A. (1956), The general circulation of the atmosphere - A numerical experiment. Quart. J. Roy. Meteor. Soc., 82: 123-164.

Rotty, R.M. (1983), Distribution of and changes in industrial carbon dioxide production. J. Geophys. Res., 88: 1301-1308.

Schlesinger, M.E. (1988), Quantitative analysis of feedbacks in climate model simulations of CO_2-induced warming. In Physically-based modelling and simulation of climate and climate change, NATO ASI Series, (M.E.Schlesinger, ed.), Part 2, Kluwer Academic Publishers, pp. 653-735.

Smagorinsky, J. (1963), General circulation experiments with the primitive equations, part I: The basic experiment. Mon. Weather Rev., 91: 99-164.

Troen, L and E.L. Petersen (1989), European Wind Atlas. Risø National Laboratory (Denmark), 656 pp.

Wiin-Nielsen, A. (1990), Observed climate variations and change - A study of the data. To be published in: The science of global environmental change (R. Corell, ed.) NATO ASI series, Springer.

Wood, F.B. (1988), Comments on the need for validation of the Jones et. al. temperature trends with respect to urban warming. Climate Change, 12: 297-311.

8.

THE WORLD'S WETLANDS UNDER THREAT
- Developing Wise Use and International Stewardship

EDWARD MALTBY
University of Exeter
Department of Geography, Amory Building,
Rennes Drive, Exeter, Devon, EX4 4RJ, England

ABSTRACT

Wetlands are posing some of today's most contentious and politically sensitive environmental questions and rival tropical rain forests for priority on the World Conservation agenda. Despite the clear recognition of the environmental importance of wetland ecosystems by early human communities there has been a progressive and devastating loss of wetland resources throughout the historic period. The consequences of loss and degradation of wetlands in the developed world may result in higher prices of goods and the need for new expensive structures to replace services previously provided free. Such options are rarely possible in the Third World where large populations are still directly dependent on wetlands. Here the consequence of wetland abuse is deprivation and hunger.

Loss of wetlands continues largely because of information deficiencies, inappropriate economic structures, deficient planning concepts and governmental policy conflict and overall institutional weakness. These issues are being addressed currently by the World Conservation Union (IUCN), many NGO's and Environment Agencies and form a focus for the environmental effort of wetland scientists.

The need for "wise use" of wetlands, a principle of the Ramsar Convention on Wetlands of International Importance Especially for Waterfowl, is paramount and should be based on sound understanding of ecosystem functioning and resulting values to society. Achieving wise use is not easy and the track record is not good. Entirely new attitudes, and new socio-economic policy and planning structures will be required. Much will depend on the ability of wetland scientists, in collaboration with other professionals, to demonstrate the real values of wellmanaged wetland resources expressed in terms in which society can appreciate the benefits arising from their conservation.

Key Words: *Ecosystem Functioning, Management, Threats, Values, Wetlands, Wise Use.*

INTRODUCTION

Questions concerning the conservation, management and wise use of wetland ecosystems are exposing some of the most contentious environmental as well as socially, economically and politically sensitive issues worldwide. The United States Government is currently suing the South Florida Water Management District over its failure to prevent degradation of wetland ecosystems in the Loxahatchee National Wildlife Refuge and the Everglades National Park caused by polluted waters released from the agricultural lands which now occupy large areas of drained marsh. Several million dollars already have been spent in legal fees and the cost of environmental restoration schemes proposed to date is approximately 500 million dollars. In June 1990, Florida's Governor Martinez signed legislation authorizing the sale of $3 billion of state bonds to acquire wetlands (Wall Street Journal, 5 July 1990). Such is the repair cost and overt official recognition of misguided past water management policies.

     This type of solution may be within the economic capabilities of wealthy nations but is not a realistic option for Third World countries wrestling with the twin problems of mounting debt and increasing pressures for development. Whilst historically society in the developed world has attempted progressively to detach itself from direct dependence on the goods and services provided by wetlands, there is by contrast a strong and often vital linkage between human communities and wetlands throughout the Third World with far-reaching implications. Ethnic violence triggered by a dispute over floodplain pastures in the Senegal Valley resulted in over a thousand deaths and tens of thousands left homeless (Africa Confidential, 1989). It is a salutary reminder of the essential coupling between sound wetland management and human welfare and of the severe repercussions when the linkage is broken. Well-managed, wetlands are among the worlds most productive ecosystems and can provide the opportunity for sustainable development, helping to meet the needs for improved living standards and increasing population pressures in the Third World. Degradation and loss is likely to exacerbate the pressures on the rural poor of the developing nations whilst reducing the options available for maintenance of environmental quality.

Never has there been such strong political rhetoric to promote sound environmental management of the worlds natural resources. Yet loss of wetlands still continues without informed public or political support. The present paper examines the case for stewardship of the world's wetland ecosystems, some of the impediments to achieving it and the initiatives which are beginning to make progress.

WETLANDS, THE WETLAND SCIENTIST AND THE ENVIRONMENTAL CHALLENGE

Wetlands cover an estimated 6 per cent of the world's land surface (Maltby & Turner, 1983) but prominence in the transitional zones between terrestrial and aquatic ecosystems complicates precise boundary delineation and area calculations. Definitions are varied reflecting in part the difficulty in giving precision to ecosystems which are themselves so varied in time and space (Maltby, 1986; Plates 1a-f; Table 1).

This has often necessitated specialisation among scientists in particular systems-peatlands, mangroves, coastal lagoons, salt marsh, freshwater marshes, swamp forest and so on - with an understandable reluctance by experts to range widely geographically and venture into unfamiliar ecosystem types. Unlike the United States where the Society of Wetland Scientists has a rapidly expanding membership currently exceeding 1,000 individuals, there is no generally recognised profession of 'wetland scientist' in Europe. Instead those who have elected to study wetlands in natural systems have emerged from the disciplines of geography, geology, hydrology, soil science, agronomy and the biological sciences. The result is a wide range of perspective and emphasis of the roles of wetlands in the landscape. The avian ecologist may see wetlands as important feeding, breeding or resting grounds for migratory birds whilst the aquatic biologist may view them as nurseries and vital habitat for freshwater, marine fish and shellfish. The hydrologist sees them as regulators of the water cycle, influencing both quality as well as quantity.

Solutions to the problems associated with sound environmental management of such diverse ecosystems will be found only by the interdisciplinary and international collaboration of wetland professionals not only representing a wide range of scientific skills but also expertise in other areas which include environmental economics, social behaviour, land and habitat management, legislation and politics. *"Increasingly....their future seems to depend on trends in economic, social and political development and the outcome of litigation, legislative and administrative debate rather than on any process in the natural world" (Maltby, 1990)*. The major environmental challenge to the wetland scientist is to interpret the wetland ecosystem as a functioning unit within a human environment which includes activities both outside and within their boundary competing for water and space, socio-economic aspirations not necessarily conducive to conservation of natural resources and institutional mechanisms which promote their degradation or loss.

Plate 1a. Everglades marsh dominated by *Cladium jamaicense* with *Typha domingensis* (cattail) in foreground. (E. Maltby).

Plate 1b. Mangrove fringe, Black River, Jamaica. (E. Maltby).

Plate 1c. *Taxodium distichum* (bald cypress) swamp, Louisiana. (E. Maltby).

Plate 1d. Groundwater supported marsh complex of Coto Donana, Spain. (E. Maltby).

Plate 1e. Aquaculture development in coastal lagoon, Messolonghi, W. Greece. (E. Maltby).

Plate 1f. Peat dominated muskeg complex, Alaska. (E. Maltby).

Table 1
Wetland classification. (Source: Dugan, 1990, modified after Scott (1989a).

1. Salt Water

1.1 Marine	1. Subtidal	I.	Permanent, unvegetated shallow waters of less than 6m depth at low tide, including sea bays, straits.
		II.	Subtidal aquatic vegetation, including kelp beds, sea grasses, tropical marine meadows.
		III.	Coral reefs.
	2. Intertidal	I.	Rocky marine shores, including cliffs and rocky shores.
		II.	Shores of mobile stones and shingle.
		III.	Intertidal mobile unvegetated mud, sand or salt flats.
		IV.	Intertidal vegetated sediments, including salt marshes and mangroves, on sheltered coasts.
1.2 Estuarine	1. Subtidal	I.	Estuarine waters; permanent waters of estuaries and estuarine systems of deltas.
	2. Intertidal	I.	Intertidal mud, sand or salt flats, with limited vegetation.
		II.	Intertidal marshes, including salt-marshes, salt meadows, saltings, raised salt marshes, tidal brackish and freshwater marshes.
		III.	Intertidal forested wetlands, including mangrove swamp, nipa swamp, tidal freshwater swamp forest.
1.3 Lagoonar		I.	Brackish to saline lagoons with one or more relatively narrow connections with the sea.
1.4 Salt lake		I.	Permanent and seasonal, brackish, saline or alkaline lakes, flats and marshes.

2. Freshwater

2.1 Riverine	1. Perennial	I.	Permanent rivers and streams, including waterfalls.
		II.	Inland deltas.
	2. Temp.	I.	Seasonal and irregular rivers and streams.
		II.	Riverine floodplains, including river flats, flooded river basins, seasonally flooded grassland.

Table 1, continued.

2. Freshwater

2.2 Lacust-
rine   1. Perman. I.   Permanent freshwater lakes (>8 ha), including
                        shores subject to seasonal or irregular inundation.
                  II.   Permanent freshwater ponds (<8 ha).
             2. Seasonal I. Seasonal freshwater lakes (>8 ha), including
                        floodplain lakes.

2.3 Palust-
rine   1. Emergent I.   Permanent freshwater marshes and swamps on inor-
                        ganic soils, with emergent vegetation whose bases
                        lie below the water table for at least most of the
                        growing season.
                  II.   Permanent peat-forming freshwater swamps, includ-
                        ing tropical upland valley swamps dominated by
                        Papyrus or *Typha*.
                  III. Seasonal freshwater marshes on inorganic soil,
                        including sloughs, potholes, seasonally flooded
                        meadows, sedge marshes, and dambos.
                  IV.   Peatlands, including acidophilous, ombrogenous, or
                        soligenous mires covered by moss, herbs or dwarf
                        shrub vegetation, and fens of all types.
                  V.   Alpine and polar wetlands, including seasonally
                        flooded meadows moistened by temporary waters
                        from snowmelt.
                  VI.   Freshwater springs and oases with surrounding
                        vegetation.
                  VII. Volcanic fumaroles continually moistened by emerg-
                        ing and condensing water vapour.
             2. Forested I. Shrub swamps, including shrub-dominated fresh-
                        water marsh, shrub carr and thickets, on inorganic
                        soils.
                  II.   Freshwater swamp forest, including seasonally
                        flooded forest, wooded swamps on inorganic soils.
                  III. Forested peatlands, including peat swamp forest.

Table 1, continued.

3. Man-made Wetlands

3.1 Aquaculture/
   Mariculture   I.   Aquaculture ponds, including fish ponds and shrimp ponds.

3.2 Agriculture   I.   Ponds, including farm ponds, stock ponds, small tanks.
                  II. Irrigated land and irrigation channels, including rice fields, canals and ditches.
                  III. Seasonally flooded arable land.

3.3 Salt Exploitation   I.   Salt pans and salines.

3.4 Urban/Industrial   I.   Excavations, including gravel pits, borrow pits and mining pools.
                  II. Wastewater treatment areas, including sewage farms, settling ponds and oxidation basins.

3.5 Water-storage areas   I.   Reservoirs holding water for irrigation and/or human consumption with a pattern of gradual, seasonal, draw down of water level.
                  II. Hydro-dams with regular fluctuations in water level on a weekly or monthly basis.

HISTORICAL IMAGES

Prehistoric communities around postglacial lake margins and the coastline of Europe depended on wetlands for food, water and materials for building, shelter and clothing (Coles & Coles, 1989). The regularly inundated, fertile floodplain environment of the Nile, Tigris and Euphrates was an essential element in the development of early civilizations. The wetlands of the Nile, Niger, Indus and Mekong valleys continue to provide for the health, welfare and safety of the people living within and close to them.

The utilization of wetlands since prehistoric times can be likened to the *"passing frontier of nature replaced by a permanently and sufficiently expanding frontier of technology"* - an attitude which *"has the recklessness of an optimism that has become habitual but which is residual from the brave days when north European free-booters overran the world and put it under tribute. We have not yet learned the difference between yield and loot. We do not like to be economic realists" (Sauer, 1938).*

The benefits appreciated by early human cultures were ignored or dismissed as insignificant during the historical period as technology enabled the rapid and progressive destruction or alteration of wetlands. The scale of loss has been immense. The coterminous United States had lost by the mid 1970's an estimated 54 per cent of its original wetland area amounting to 87 million hectares (Tiner, 1984). The national loss rate for wetlands between the 1950's and the 1970's averaged 185,000 ha per year (Tiner, 1984).

Rates of loss are less well documented in Europe but their proportions are almost certainly higher given the longer period of human intervention and generally higher population densities. About half the Netherlands, once part of the complex delta of the Rhine, Meuse, Ems and Scheldt rivers, would be regularly inundated were it not for dams, dykes and engineering works built since the 8th Century. Wetland loss has been viewed as a largely "progressive, publicspirited endeavour" into the present period (Baldock, 1984). New drainage for agriculture has continued throughout Europe despite food overproduction, price support and various forms of subsidy. 40 per cent of the coastal wetlands of Brittany have been lost since 1960 and two-thirds of the remainder are seriously affected by drainage and similar activities (Mermet, in Baldock, 1984). In Portugal about 70 per cent of the wetlands of the western Algarve have been altered for agricultural and industrial development (Pullan, 1988). Drainage in New Zealand continues to reduce the wetland area which already is less than 10 per cent of that present at the time of original European settlement (Smith, 1986). 50 per cent of Irish peatland have been lost as natural ecosystems (Van Eck et al., 1984) whilst between the middle of the nineteenth century and 1978 84 per cent of lowland raised bog in Britain was lost through afforestation, agricultural reclamation and commercial peat cutting. Only 6 per cent of the original 13,000 ha is considered intact (Nature Conservancy Council, 1986).

A notable change in attitude has taken place recently in Denmark with public support for restoration of wetlands rather than their drainage. All wetlands over 1 ha in size are now protected by law and no change in ecological state can be made without a special permit.

WETLAND FUNCTIONING

The importance of wetlands is based on the goods and services significant for:

I.   Direct and indirect human use
II.   Welfare of wildlife and maintenance of the gene pool
III.   Environmental maintenance and quality.

These roles were implicit in the existence of early European communities and are the key to survival of large Third World populations today.

Hydrological, biological, chemical and physical processes occurring naturally in wetlands result in functions such as groundwater recharge, flood control and nutrient transformation. Process interactions maintain ecosystem elements or components which in turn can generate products such as forest, wildlife, fisheries and grazing resources. At the ecosystem level attributes include biological diversity and cultural uniqueness or heritage. All three of these aspects of ecosystem dynamics may result in sustainable benefits to human society whether directly or indirectly. However wetlands vary in process dynamics and this results in significant variation in functional characteristics (Table 2).

Sound environmental management depends on accurate recognition of how particular wetlands function and an understanding of the ways in which functions, products and attributes are impacted by human activities. It is especially important to identify levels of tolerance to adverse impacts and in the Third World means of enhancing wetlands to generate optimum levels of products on a sustainable basis without compromising other important values.

The emphasis previously in Europe has been on more traditional nature conservation values such as uniqueness, representativeness and rarity. In the United States, however, significant recent progress has switched the emphasis more to functional assessment methods for the evaluation of wetlands. Adamus and Stockwell (1983) identify a wide range of biophysical characteristics from which to predict the likelihood that a wetland would perform a particular function. Originally designed to provide guidance to government agencies in planning the most environmentally suitable alignment for new roads the Adamus approach has become an important tool for wetland conservation in the United States. It has formed the basis of the Wetland Evaluation Technique evolving under the guidance of the U.S. Corps of Engineers and Environmental Protection Agency, the agencies jointly responsible for wetland regulation and permitting in the USA (Adamus et al., 1987)

A number of examples will serve to illustrate the environmental and ecological importance of wetlands.

FLOOD CONTROL

Wetlands store and detain precipitation and runoff thus reducing flood peaks. This natural storage can reduce the necessity for expensive dams and defensive engineering structures. Evidence from Wisconsin suggests that catchments with 15 per cent wetlands and lakes had flood peaks 60-65 per cent lower than they would be without them (Adamus & Stockwell, 1983). Wetlands on the Charles River in Massachusetts have been preserved as natural flood defences.It was estimated that if they had been completely filled the increase in flood damage would have exceeded $17 million per year (U.S. Corps of Engineers, 1972), putting a flood prevention value of some $13,500

Table 2
Wetland values. (After Dugan, 1990).

	Estuaries without mangroves	Mangroves	Open coasts	Flood plains	Freshwater marshes	Lakes	Peatlands	Swamp forests
Functions								
1. Groundwater recharge	○	○	○	■	■	■	●	●
2. Groundwater discharge	●	●	●	●	■	●	●	■
3. Flood control	●	■	○	■	■	■	●	■
4. Shoreline stabilisation/ Erosion control	●	■	●	●	■	○	○	○
5. Sediment/ Toxicant retention	●	■	●	■	■	■	■	■
6. Nutrient retention	●	■	●	■	■	●	■	■
7. Biomass export	●	■	●	■	●	●	○	●
8. Storm protection/ Windbreak	●	■	●	○	○	○	○	●
9. Micro-climate stabilisation	○	●	○	●	●	●	○	●
10. Water transport	●	●	○	●	○	●	○	○
11. Recreation/Tourism	●	●	■	●	●	●	●	●
Products								
1. Forest resources	○	■	○	●	○	○	○	■
2. Wildlife resources	■	●	●	■	■	●	●	●
3. Fisheries	■	■	●	■	■	■	○	●
4. Forage resources	●	●	○	■	■	○	○	○
5. Agricult. resources	○	○	○	■	●	●	●	○
6. Water supply	○	○	○	●	●	■	●	●
Attributes								
1. Biological diversity	■	●	●	■	●	■	●	●
2. Uniqueness to culture/ Heritage	●	●	●	●	●	●	●	●

Key: ○ = Absent or exceptional; ● = Present; ■ = common and important value of that wetland type.

per hectare per year on the intact wetland area. Plans to build expensive artificial structures were shelved when it was realised that the role was filled less expensively by the natural system. This contrasts with the decision to build the Danish Wadden Sea seawall which now causes increased erosion in other parts of the Wadden Sea coastline (J. Fjeldså, pers.comm.).

This least cost solution to flood prevention is particularly appropriate to the needs of the worlds poorest nations but is often counter to some of the arguments raised in support of large dams and extensive levees. It is noteworthy that recent opposition to a wetland drainage programme in the Dong Thap Province of the Mekong Delta, Vietnam was based partly on flood storage considerations (Scott, 1989b).

RETENTION AND TRANSFORMATION OF SEDIMENT, CONTAMINANTS AND NUTRIENTS

Regular deposition of nutrient-rich sediment contributed to the success of early agriculture along rivers such as the Nile, but in addition to fertility such sediments are also important in the maintenance of the physical stability of the floodplain and delta environments. It is only when impeded that the full value of the function can be demonstrated. Overbank flooding and sediment deposition in the Mississippi delta has been dramatically reduced from historical levels by levée construction and channelisation. Dams upstream have further reduced the amount of sediment carried by the river by some 50 per cent (Keown et al., 1980). Most of the load which is transported is now delivered directly to the Gulf of Mexico and beyond the continental shelf. It is argued that this, together with an increasing density of man-made canals, exacerbate the rate of land loss in the delta which may exceed 0.5 per cent per year as deposition fails to match the rate of tectonic subsidence and sea level rise (Scaife et al., 1983; Turner et al., 1982; Mendelssohn et al., 1984). In the case of the Nile Delta recent estimates indicate that the coastline is receding at up to 16-30 m per year as sharply reduced sediment input because of the Aswan High Dam fails to offset continued subsidence and eustatic sea level rise (Stanley, 1988).

Various biological, chemical and physical processes, many still imperfectly understood, immobilise and transform a wide range of environmental contaminants and nutrients which in excess cause major ecological problems. Wetlands may remove heavy metals, pesticides or industrial residues and other toxicants generally in forms more or less tightly bound to sediment or soil particles. Large amounts of phosphorus including fertilizer runoff can be inactivated by chemical bonding to inorganic ions (Richardson, 1985) thus preventing harmful eutrophication of adjacent open waters. Of particular importance, however, is denitrification which can remove in gaseous form 40-98 per cent of the nitrogen from a wetland system (Adamus & Stockwell, 1983). The process is favoured by alternating conditions of flooding and drainage or other conditions which create the close temporal and/or spatial juxtaposition of anaerobic and aerobic micro-environments. Under aerobic conditions NH_4+ released by the breakdown of organic matter, is converted to nitrate (NO_3-). The nitrate can be transformed under anaerobic conditions to molecular dinitrogen and gaseous nitrous oxide by the process called denitrification. These gaseous products are immediately unavailable to plants and micro-organisms and are lost from the system. The wetland thus acts as a sink and prevents nitrate build up which in excessive quantities can cause major problems of eutrophication. Nitrate runoff from fertilized agricultural areas is transformed in exactly the same way.

Wetlands do the job of preventing nitrates leaking into freshwaters naturally and at no direct cost. Their destruction removes this benefit, a fact realised increasingly as nitrate levels in streams and groundwaters have increased generally throughout Europe and North America. An environment rich in wetlands can at least reduce the potentially harmful impacts of current or past fertilizer use in the landscape. Recent estimates from Sweden indicate that a wetland area of 2 km_2 would reduce nitrogen inputs into adjacent waters by nearly 2,000 tons a year (Fleicher, 1990).

There is now considerable interest in reconstructing wetlands or rehabilitating degraded ecosystems to restore functions of nutrient and contaminant retention. Both capital and maintenance costs may be substantially less than the artificial systems required to recover similar levels of water quality. Considerably more scientific research is required to establish the optimum environmental conditions.

Odum and Ewel (1974) have proposed the use of cypress swamps in Florida as natural tertiary treatment centres for domestic waste water. They found that 98 per cent of all nitrogen (N) and 97 per cent of all phosphorous (P) was removed before waste water entered the groundwater. It is unlikely however that sufficiently large areas of natural wetlands exist in close proximity to centres of population in the developed world which would remain ecologically resilient and could be used for this purpose. However the situation is quite different in developing countries where aquatic plant species such as water hyacinth (*Eichornia crassipes*) are particularly effective in absorbing contaminants and nutrients. The plant can remove 92 per cent of N, 60 per cent of P and 97 per cent of BOD in seven days exposure. Although the plant itself can become a nuisance because it spreads rapidly over open water, restricts navigation and reduces dissolved oxygen levels, current research is aimed at utilising ancillary benefits such as fertilizer, biogas and animal feed which could be obtained from its harvest (Thyagarajam, 1983)

CARBON DYNAMICS

The development of organic soils and peat in wetlands, a consequence generally of waterlogged conditions reducing decomposition but which may be aided further by acidity, nutrient deficiency and/or low temperatures, represents one of the largest carbon (C) pools in the terrestrial biosphere. Peatlands occupy just 3 per cent of the worlds land area yet they store almost 20 per cent of the globe's soil carbon pool (Maltby et al., in prep.). The role of wetlands as a net carbon sink with links in the biogeochemical cycling of CO_2, CH_4, H_2S and N_2O has been examined by Armentano and Menges (1986). With the exception of H_2S these gases absorb infra-red energy and changes in their concentration in the atmosphere may influence the earth's radiation balance and the global climate.

Armentano and Menges (1986) have demonstrated a significant shift in the balance of carbon movement between wetlands and the atmosphere as a result of human intervention, particularly drainage for agriculture. It is estimated that before substantial human disturbance, wetlands were a sink for 200 million t carbon/a, equivalent to about 4 per cent of fossil fuel emissions today. Historically, 6.5 billion t carbon may have been displaced by 1980 (Armentano and Menges, 1986). They found differential shifts according to region, some areas of the world acting as reduced sinks but others actually converted to net sources of carbon. By 1980 the net carbon sink in Finland and the USSR had reduced by 21-33 per cent, in western Europe by almost half whilst in central Europe the sink had been lost completely. Oxidation of previously accumulated organic carbon is accelerated in high latitudes by peat extraction for fuel and horticultural use whilst in the peat mangrove swamps of the tropics by excavation for aquaculture ponds. Intact wetlands naturally generate methane which is on a molar basis 3.7 times (10 times on a weight basis) more potent as a greenhouse gas compared with CO_2 (Lashof and Ahuja, 1990). These emissions have been estimated at 110 million tons a year, equivalent to 82.5 million tons C a year (Matthews and Fung, 1987). It is still not clear just how significant is the increased release of CO_2 compared with possible decreases in CH_4-emissions from intact wetlands. However, the possible role of wetlands in helping to maintain global climatic stability cannot be ignored. It exemplifies wetland ecosystems as part of the global commons-enhanced carbon dioxide release from any part of the world may have repercussions well beyond the boundaries of the disrupted wetlands actually contributing the carbon. The case is also a reminder of the potential of a major cumulative impact that can arise from many separate decisions to exploit a natural resource but which individually might have negligible effect on biogeochemical cycling. The assessment of the relative importance of wetlands as a climate stabilizer should be a high priority in future research programmes.

COMPONENTS AND PRODUCTS

The plants and animals that live in or depend on wetlands for at least part of their life cycle may yield harvests or products of great value and benefit to human society. The swamp sago (*Metroxylon sagu*), an important component of the floodplain swamps of SE Asia provides the main food staple for a quarter of the population of Irian Jaya (Koonlin, 1980). Exploitation of mangrove products including timber, fuelwood, charcoal and tannins earn hundreds of millions of export dollars as well as yielding materials for domestic consumption (Hamilton and Snedaker, 1984). Many of the coastal wetlands that remain in NW Europe provide important grazing areas and sheep utilizing salt marshes can often demand a market premium. Over half a million wild herbivores, mainly antelope, live on the Sudd floodplains in Africa and hunting provides up to 25 per cent of the annual meat intake of the local people (Drijver and Marchand, 1985). Submergent plant species including *Myriophyllum* and *Elodea* are harvested on the Andean Altiplano to support cattle in the dry season, providing a valuable resource at a time when other feed is in short supply (J. Fjeldså, pers. comm.).

FISHERIES SUPPORT

Both freshwater as well as coastal wetlands provide fish with nutrient rich protective habitats especially important for feeding, spawning and as nursery areas. It is estimated that two thirds of the fish we eat depend on wetlands at some stage in their life cycle. Ninety per cent of the fish harvest of the Gulf of Mexico comprises wetland dependent species. The coastal flats of the Wadden Sea in Europe supports about 60 per cent of the North Sea's brown shrimp, over 50 per cent of the sole, 80 per cent of the plaice and almost all of the herring stock for some part of their life cycle. Over 25,000 villagers around lake Chilwa on the Malawi-Mozambique border rely on the lake-edge fishery. Here, as with lake and floodplain environments throughout the world, animal migration and spawning of fish in the marginal wetlands depends on the flood cycle. Its success is vital to the survival of the fishing communities. In the lower Mekong Basin 236,000 out of a total catch of 500,000 tons is estimated to be derived from wetlands. In 1981 the fisheries value of wetlands alone contributed $90 million to the economy and supplied 50-70 per cent of the protein needs of the Mekong delta's 20 million people (Pantulu, 1981).

Reduction of coastal wetland area results in a decline in commercial fisheries (Turner, 1973; Maltby et al., 1988). Interruption or alteration of flood regime similarly can have major repercussions for downstream fish stocks. Nowhere has this been better demonstrated than in the case of the closure of the Aswan High Dam in Egypt. The flow of the Nile was reduced by 40 km^3 between 1962 and 1976, reducing significantly the input of sediment and nutrients to coastal waters of the Eastern Mediterranean. Primary production fell and despite an increase in fishing effort by a factor of nineteen from 1952 to 1962 the Egyptian catches dropped from nearly 38,000 tons to little more than 7,000 tons. The decline was particularly sharp in sardine stocks, resulting in the closure of two coastal canning factories (Rzoska, 1976).

In the developed world resource reduction can be offset in part by market forces raising prices or expensive structures replacing natural functions. Such options among the rural poor of the Third World rarely exist and the consequence of functional disruption is deprivation and hunger.

ATTRIBUTES

Wetland ecosystems have special asset significance resulting from biological diversity and recognition of cultural and heritage values. They provide habitat for a rich and diverse assemblage of plants and animals many of which are listed as rare or endangered. Only a fraction of their genetic resource has been studied and an even smaller proportion tapped for direct human use. The economic and social importance of rice, sago, oil palm, shrimp, oysters and waterfowl underline the practical significance of the resource. Yet even more important genetic material may find economic application in the future - perhaps salt marsh species adapted to utilise soils damaged by salinization or the physiological mechanisms of insectivorous wetland plants to improve production in nitrogen deficient environments. It is impossible to express in exact terms what economists term the "inheritance value" of wetlands but past experience urges us to ignore it at our peril.

Wetlands are important for the support of endemic and frequently rare species. In Lake Tanganyika alone 214 species of Cichlid fish have been identified of which 80 per cent are endemic (Moss, 1980). Inaccessibility of wetlands has added to their importance in maintaining some species. Thus the mangrove forest of the Sundarbans in India and Bangladesh is the largest remaining habitat for the Bengal Tiger, *Panthera tigris*.

The Florida panther, *Felix concolor coryi* would be extinct without the remaining large tracts of wetland in Southern Florida but habitat fragmentation by roads and other development continues to threaten the small residual population of 20-50 individuals (Noss, 1987).

Spectacular landscapes and especially concentration of wildlife are much valued features of wetlands. Their role as wintering, breeding or feeding staging posts along extensive migratory routeways makes this a particular characteristic of waterfowl. In Mauritania the tidal flats of the Banc d'Arguin National Park provides a wintering site for 3 million shorebirds (Dugan, 1990). The floodplains of the Senegal, Niger and Chad basins support over a million waterfowl (Monval et al., 1987) many of which will spend the summer in the Arctic, northern or western Europe.

These international linkages between wetlands create a dimension to wetland management and conservation which was recognised by the original architects of the Convention on Wetlands of International Importance Especially as Waterfowl Habitat signed at Ramsar, Iran in 1971. The Convention has become the major intergovernmental forum for the promotion of wetland conservation. However the current membership of sixty two contracting parties represents but a third of the worlds nations and much needs to be done to improve coverage in the Third World.

There is increasing recognition of the multiple benefits to be obtained from maintenance of the functional integrity of wetlands. Given sensitive management a wide range of products and services can be provided at no or little cost to the environment. The extent to which large human populations can be supported and benefit from sound management strategies is well summarised for the tropics in Dugan, (1990).

Despite this growing perception, however, wetland degradation and loss continues with major environmental consequences.

OVERRIDING FACTORS PROMOTING WETLAND DEGRADATION AND LOSS

The reasons why people have continued to drain and convert wetlands to nonwetland uses or carry out activities incompatible with maintenance of their functional integrity are examined in Maltby (1986, 1988) and Dugan (1990). The salient issues are complex but are linked invariably to basic information deficiencies, inappropriate economic structures, deficient planning concepts and governmental policy conflict and overall institutional weaknesses.

1. Science and Information Base
Politicians, decision-makers, farmers and the public are generally unaware of or choose to ignore the real value of the environmental work done by wetlands. This is partly because services like water quality maintenance is a free product of the wetland ecosystem. The costs of replacing such natural benefits rarely features in the equation deciding the fate of a wetland area. The recent review of current issues and required action in wetland conservation identifies the limited appreciation of the importance of wetlands as of major significance (Dugan, 1990). The lack of a systematic effort to measure the non-market values of wetlands and an inability to relate these to decision-makers have enabled wetland degradation to proceed unchallenged on the terms on which development decisions normally are taken. Farmers have often not drained wetlands out of greed or malicious motives but because of inbuilt utilitarian views of the environment and lack of knowledge of the real values of the wetland. Mechanisms of transfer of the appropriate information about wetland ecosystems lack the efficiency and effectiveness to ensure that the most relevant people gain the most appropriate knowledge.

2. Imbalance of Costs and Benefits
The farmer who invests in a wetland drainage project for agriculture may (but not necessarily) reap considerable financial benefits against his capital risk. However, the environmental costs of his project may have to be shared by others with no say at all in the initial decision. These costs, such as deterioration of water quality, reduction in fisheries and loss of wildlife habitat, may arise at a considerable distance from the point of impact and may be well beyond the boundary of the area immediately affected. Conversely the landowner himself may not obtain a significant share of the public or downstream benefits which result from wetland conservation yet the costs of sound management to sustain public benefits would often fall to the private landowner. New planning strategies must be designed to take account of the uneven distribution of the costs and benefits of wetland maintenance or alternative land management

systems. Grants and subsidy mechanisms must take this inequality more fully into account. It is clear that a holistic planning solution must be developed to enable a more appropriate balance to be obtained in the distribution of costs and benefits to society of maintaining wetlands within in the landscape.

3. Policy Conflicts

Governments and supragovernmental bodies (such as the European Commission) are organised sectorally. This often results in inherent conflicts of aims and objectives. The aims of a national agricultural policy may include improved crop and animal production and the welfare of farmers achieved traditionally by means including wetland drainage and intensification of land use. The environment agency of the same government meanwhile may have a clear mandate to promote wetland conservation and oppose such strategies.

A wide range of policies have had unintended effects on wetland ecosystems particularly in the developed world. Grants, subsidies, artificial price support and tax incentives are all mechanisms which have favoured policies resulting in wetland degradation and loss throughout Europe and North America. The actual costs of drainage which are often subsidised from public funds though artificially high crop prices may in themselves be sufficient inducement for farmers to convert wetlands. Wheat prices in the European community 1978-80 were 40-60 per cent above the world market level. This undoubtedly contributed to the conversion of lowland wet grazing meadows to winter wheat production (Dugan, 1990). Sugar prices in the United States are held significantly above the world price so adding to the profitability of converting Florida wetlands for production of sugar cane. The development of processing facilities, establishment of associated employment and infrastructure all makes reversal of such ill-advised policy decisions difficult on social grounds.

Soft loans, grants, aid packages and world economic conditions may all promote wetland loss throughout the developing world despite increasingly overt requirements by donor agencies for environmental sensitivity. Third World nations are caught in the dilemma of development for at least short term economic and social gains, responding to the immediate needs of the population versus wetland resource management for the longer term, maintaining ecological and environmental, as well as economic, options for the future. "We cannot afford not to develop" is the common argument used not only by politicians but also by environmentalists in poor countries. This creates pressures for the intentional alteration of wetland ecosystems. One of the most recent potential developments in the tropics is the mining of peat for energy. The opportunities provided to reduce fuel import bills and balance of payments deficits, establish local industry, reduce fuel wood losses and raise standards of living are very tempting. Resurgence in the world oil price in the second half of 1990 will undoubtedly increase the pressure on tropical peatlands in Jamaica, Indonesia, Africa and South America. The environmental implications of a peat mining for energy policy are immense but understanding of the adverse impacts has lagged significantly behind the interest in development (Maltby, 1988).

4. Institutional Deficiencies

Politicians or the governments they represent are rarely gripped by the subject of wetlands. It is unusual for the management problems they pose to be given the level of attention and resources for action which are required. Institutional shortcomings include lack of integration among different responsible agencies, inadequate management systems, shortage of experienced and suitably trained professionals, inadequate and/or non-enforced legislation, and insufficient funding levels. Until institutional weakness is overcome by popular support for and recognition of the full environmental role of wetlands, these ecosystems will continue to disappear.

THE WAY AHEAD

The magnitude of the task to change the attitudes and establish the mechanisms for the sound environmental management of wetland resources should not be under-estimated. It must start with the more general recognition of wetland science as a discipline of professional standing which can command the attention of decision-makers. This will require appropriate course structures and practical training. Universities, educational funding agencies and other interested parties must take initiatives in the development of suitable courses leading to professional qualifications in wetland science and management. Appreciation of the nature and importance of wetlands needs to be strengthened throughout the school curriculum particularly at the primary level. In this way information about wetlands becomes part of an individuals knowledge base before he or she becomes an engineer, farmer, politician, householder or nature manager and develops more specialised and narrower perspectives on the use of land resources.

Appreciation of and concern for the "common good" is a challenge to society as a whole which will be tested increasingly by the need to respond to major global issues such as climatic change, sea level rise and pollution. It is no less a requirement to ensure the sound management of wetland resources. Where such recognition may be deficient for cultural or other reasons it is unrealistic to imagine how such an attitude could be developed other than gradually and through a well informed educational system supported by appropriate media attention on information transfer.

The World Conservation Union (IUCN) has identified twelve priorities for action to strengthen wetland conservation:

1.   Support national and regional wetland conservation programmes
2.   Improve the quality and quantity of information on national wetland ecosystems
3.   Develop national policies which support wetland conservation and promote appropriate legislation
4.   Improve methodologies for planning use of wetland ecosystems
5.   Support conservation of critical wetland habitats

6.   Develop tools for wetland conservation that contribute to sustainable development
7.   Strengthen wetland management institutes
8.   Widen the acceptance of the principles and concepts of wetland conservation by decision-makers in government and in the public and private sectors
9.   Strengthen international collaboration for wetland conservation
10.   Support wetland conservation through development assistance
11.   Enhance cooperation between international institutions working on wetland conservation and related issues
12.   Identify and develop action to address critical issues of the 21th century.

These priorities, expanded in Dugan (1990), should be centred around National Wetland Programmes and supported by the collaboration of government agencies, NGO's, scientists, economists and other appropriate specialisms. They are consistent with the concept of "wise use" recognised increasingly as a fundamental tenet of the Ramsar Convention. The Third Meeting of the contracting parties proposed the following definition:
       "The wise use of wetlands is their sustainable utilisation for the benefit of humankind while maintaining the natural properties of the ecosystem."Sustainable utilisation is "human use of a wetland so that it may yield the greatest continuous benefit to present generations while maintaining its potential to meet the needs and aspirations of future generations."
       The concept of wise use is crucial to the implementation and wider acceptance of the relevance of the Ramsar Convention particularly among the world's developing nations. Here there may be limited support for the sentimental, aesthetics or gene pool values of migratory birds but the direct welfare of communities is of overriding significance. Fortunately, what may be good for ecosystem health in supporting human needs is often also what is required to support wildlife. There is still, however, a wide gap between defining wise use and applying its principles to specific wetlands or countries. Identification of acceptable patterns and intensity of use is often difficult and depends on a wide range of factors including wetland type, regional context, existing use by human communities and their future needs. The wetland scientist will be only one member of the team required to devise management plans based on the wise use concept.
       In general terms wise use will require

-   identification of wetland functions, components, attributes and values
-   integration of compatible uses where possible
-   separation of incompatible uses
-   zoning and environmental planning
-   catchment (river basin) management
-   appropriate employment/social/economic strategies to relieve the ecosystem of damaging human pressures.

Suitable methods of assessing wetland functions are high on the scientific agenda. The Wetland Evaluation Technique developed in the United States (Adamus et al., 1987) is appropriate to support a well developed regulatory programme but does not translate to the non-regulated European environment. It is generally too sophisticated for most Third World needs. A major EC programme has been initiated which aims to develop a methodology for the functional analysis of European Wetland Ecosystems (Maltby and Barth, in press) and IUCN has promoted the development of rapid assessment techniques which can be used in the tropics (Barbier, 1989). Development of wise use depends on the active participation of local people using appropriate traditional or other techniques consistent with management requirements. It depends also on decision-makers at all levels in government and outside associated with development aid and related projects recognising the need to maintain essential ecological and environmental processes. Stewardship of the world's remaining wetland resources, particularly throughout the Third World is not the preserve of local populations but a requirement for a suitably informed population worldwide.

Even when accepted, wise use may be inhibited by processes and events well beyond the immediate boundaries of the wetland. This is a particular problem associated with water abstraction and regulation. It is futile to conserve, manage or use wisely a wetland if circumstances elsewhere in the catchment cause hydrological changes which degrade the ecosystem. Thus in Spain, the Tablas de Damiel National Park and a designated Ramsar site has been seriously altered as a result of water table lowering by groundwater abstraction for irrigation on agricultural land outside the park (Llamas, 1988). Reduced freshwater flows in Lake Ichkeul, Tunisia, caused by the damming of headwaters a considerable distance upstream of the coastal wetland is resulting in large salinity changes and loss of biological diversity (Hollis, 1986). Plans to divert the flow of the Acheloos river in western Greece may have serious repercussions for the remaining wetland complex at Messolonghi. Drainage of former Lake Karla in Greece has resulted in so much pollution that the inhabitants of Volos, some fifteen kilometres distant, have rioted in the streets. Similar civil disorder has occurred in Huelva, Spain because of the contamination of shellfish by pollution from upstream mining activities (Maltby et al., in press).

The only way in which such environmental disruption can be avoided is through the development of integrated management structures for the entire catchment area of the wetland ecosystem. Progress in meeting this need has been very limited to date and will depend on major reorganisation of the planning process. A recent action by the Commission of the European Communities Directorate General XI (Environment) has focused on the need for and problems of Integrated Management of Coastal Wetlands of Mediterranean Type. The conclusions of the first phase of this action are given in Table 3.

Table 3
Conclusions of the first phase of a CEC Action Preparatory to the integrated
management of coastal wetlands of Mediterranean type.

- Wetlands should be a self-regulating system, characterized by primary biotic conditions;

- No wetland area left which could be sacrificed for uses and functions other than those provided by wetlands;

- Conservation and maintenance of functions requires integrated management;

- An instrument for functional analysis based on natural and as far as possible socio-economic values is urgently required;

- Wetland inventory is required, regularly updated, accessible and at a scale appropriate to management;

- Zoning is required both of the wetland and the area outside where environment impact assessment is necessary for relevant uses to avoid wetland degradation;

- Integrated Management Plan is an essential instrument to ensure satisfactory resource conservation;

- Executive management bodies must be established to plan and enforce sustainable use;

  - must involve local people and authorities;
  - may be a particular service or institutionalised collaboration;
  - must be competent;

- Need to raise awareness of need for integrated management and train managers and people involved with issues;

- Specific guidelines for users must be developed for the wetland and its catchment especially in regard to water and pollution management;

- Controlling and combating illegal use crucial but highly sensitive in implementation of wetland conservation and integrated management.

CONCLUSIONS

Wetlands are not wastelands. These ecosystems function, possess components and have attributes which are both valuable to human society as well as vital to the maintenance of environmental quality. Yet wetland resources worldwide continue to be degraded and lost, generally on the basis of inadequately informed decisions together with a failure of existing institutional mechanisms to properly evaluate them. The wise use of wetlands for the greatest benefit to both humankind as well as other species depends on new attitudes and new socio-economic, political and planning structures. Wetlands cannot be viewed in isolation; they are parts of river basins, coastal zones and biogeochemical cycles; they are part of landscapes used by people. Such use in the Third World may be intimate and represent the basis of survival. They are parts of complex social and economic systems. The concept of wise use must include this broader prospective of wetland ecosystems.

The professional challenge to the wetland scientist is to interpret the wetland ecosystem as a functioning unit within the complex human and often dynamic natural environment, to evaluate their tolerance of various uses and advise on optimum management strategies to maintain functional integrity. His or her more difficult task is to collaborate with other professionals and lobbyists to bring about major changes in the institutional structures required at local, regional, national and international levels to ensure the future wise stewardship of some of the globe's most precious yet fragile environmental resources.

ACKNOWLEDGEMENTS

This paper has benefitted from the author's close involvement with IUCN's Wetland Advisory Committee and preparation of the document "Wetland Conservation". Alistair Maltby is thanked for his word processing of my draft.

REFERENCES

Adamus, P.R., Clairain, E.J. Jr., Smith, R.D. and Young, R.E. (1987), Wetland Evaluation Technique (WET), Vol II: Methodology. Operational Draft Report US Army Engineer Waterways Experimental Station, Vicksburg, Miss., Report Y-87.

Adamus, P.R. and Stockwell, L.T. (1983), A Method for Functional Assessment: Vol 1. Critical Review and Evaluation Concepts. FHWA - IP 82 83 US Dept. of Trans. Fed. Highway Admin., Washington, D.C.

Africa Confidential (1989), Mauritania: War on Black Citizens.
Africa Confidential 30 (14) 2-3.

Armentano, R.V. and E.S. Menges (1986), Patterns of change in the carbon balance
of organic soil - wetlands of the temperate zone.
Journal of Ecology 74, 755-774.

Baldock, D. (1984), Wetland drainage in Europe IIED/IEEP, London.

Barbier, E.B. (1989), The Economic Value of Ecosystems: 1 - Tropical
Wetlands. Gatekeeper Series No. LEEC 89-02, IIED, London.

Coles, B.J. and Coles, J.M. (1989), People of the Wetlands. Bogs, Bodies and Lake
Dwellers, Thames and Hudson.

Cowardin, L.M., V. Carter, F.C. Golet and E.T. La Roe (1979), Classification of
wetlands and deepwater habitats of the United States. U.S. Fish and Wildlife Service
Pub. FWS/OBS-79/31, Washington.

Drijver, C.A. and M. Marchand (1985), Taming the floods. Environmental Aspects
of Floodplain Development in Africa. Centre for Environmental Studies, State University of Leiden.

Dugan, P.J. (1990), Editor, Wetland Conservation. A Review of Current Issues and
Required Action. IUCN, Gland, Switzerland, 96 pp.

Fleischer, S. (1990), Wetlands - a nitrogen sink. Acid Magazine 9, 26-28.

Hamilton, L.S. and S.C. Snedaker (eds.) (1984), Handbook for Mangrove Area
Management. UNEP - East-West Centre - IUCN, Gland, Switzerland.

Hollis, G.E. (ed.) (1986), Modelling and Management of the Internationally Important
Wetlands at Geraet el Ichkeul. Tunisia IWRB, Slimbridge, UK.

Keown, M.P., E.A. Dardean and E.M. Causey (1980), Characterisation of the sus-
pended-sediment regime and bed-material gradation of the Mississippi River Basin,
U.S. Army Engineers District, New Orleans, Louisiana, Potamology Investment
Report No. 221. Vicksburg MS: Environmental Laboratory,
U.S. Army Engineers Waterways Experimental Station.

Koonlin, T. (1980), 'Logging the swamp for food'. In W.R. Stanton and
M. Flach (eds.), Sago. The Equatorial Swamp as a natural resource.
Martinus Nijhoff Publishers, Kuala Lumpur, Malaysia, 13-34.

Lashof, D.A. and Ahiya, D.R., (1990), Relative contributions of greenhouse gas emissions to global warming. Letter to Nature 344 529-531.

Llamas, M.R. (1988), Conflicts between Wetland Conservation and Groundwater Exploitation: Two Case Histories in Spain. Environ. Geol. Water Sci. 11, pp. 241-251.

Maltby, E. (1986), Waterlogged Wealth - why waste the world's wet places? Earthscan, London.

Maltby, E. (1988a), Global Wetlands - History, Current Status and Future. In Hook, D. (ed.), The Ecology and Management of Wetlands, Croom Helm, 3-14.

Maltby, E. (1988b), Peatland Development in Indonesia. Mimeo Report to IUCN, Gland, Switzerland, 30 pp.

Maltby, E. (1990), Wetland Management Goals: Wise Use and Conservation. Keynote Address, The People's Role in Wetland Management, Leiden June 1989.

Maltby, E. and Barth, H. (in press), Editors, Wetland Functioning and Values. Commission of European Communities, Brussels.

Maltby, E. and R.E. Turner (1983), Wetlands of the World, Geographical Magazine LV, 12-17.

Maltby, E., Hughes, R. and Newbold, C. (1988), The dynamics and functions of Coastal Wetlands of the Mediterranean type. Mimeo Report to CEC DG XI (Contract No. 6611/2H/10), Brussels. 112 pp. + Appendices.

Maltby, E., Immirzi, P. and Clymo, R.S. (in preparation), The role of peatlands in carbon dynamics.

Matthews, E. and Fung, I., (1987), Methane emission from natural wetlands: global distribution, area, and environmental characteristics of sources. Global, Biogeo. Cycles 1, 61-86.

Mendelssohn, I.A., R.E. Turner and K.L. McKee (1984), Louisiana's eroding coastal zone: management alternatives. Journal Limnological Society South Africa 9, 63-75.

Monval, J.Y., Pirot, J.Y. and Smart, M. (1987), Recensements d'anatides et foulques hivernaut en Afrique du Nord et de l'Ouest: janvier 1984, 1985 et 1986. IWRB, Slimbridge, UK, 44 pp.

Moss, B. (1980), Ecology of Freshwaters. Blackwell, Oxford.

Nature Conservancy Council (1986), Nature conservation and afforestation in Britain, Peterborough, England.

Noss, R.F. (1987), Protecting natural areas in fragmented landscapes. Natural Areas Journal 7, 2-13.

Odum, H.T. and K.C. Ewel (1974), Cypress Wetlands for Water Management Recycling and Conservation. Annual Report to National Science Foundation and Rockefeller Foundation, Centre for Wetlands, University of Florida, Gainesville, Florida.

Oliver, G.A., Hogan, D.V. and Maltby, E., (1990), Functional Analysis of Wetlands in Messolonghi and Odiel. Mimeo Report to CEC DGXI (Contract No. ZM/6600 (89)(ET1), Brussels. 118 pp + Appendices.

Pantulu, V.V. (1981), Effects of water resource development on wetlands in the Mekong basin. Environment Unit, Mekong Secretariat Bangkok. Unpublished Mss.

Pullan, R.A. (1988), A Survey of the Past and Present Wetlands of the Western Algarve. Department of Geography, University of Liverpool, UK, 100 pp.

Richardson, C.J. (1985), 'Mechanisms controlling phosphorus retention capacity in freshwater wetlands', Science 228, 1424-7.

Rzoska, J. (1976), A Controversy Reviewed. Nature 261, 444-5.

Sauer, C.O. (1938), Theme of Plant and Animal Destruction in Economic History. Journal of Farm Economics 20, 765-75.

Scaife, W.W., R.E. Turner and R. Costanza (1983), Recent land loss and canal impacts in coastal Louisiana. Environmental Management 7, 433-442.

Scott, D.A. (1989a), Design of Wetland Data Sheet for Database on Ramsar Sites. Mimeo to Ramsar Convention Bureau, Gland, Switzerland, 41 pp.

Scott. D.A. (1989b), A Directory of Asian Wetlands, IUCN, Gland, Switzerland and Cambridge, UK, 1181 pp.

Smith, K. (1986), Pulling the plug on West Coast Wetlands. Forest and Bird 17(1), 12-12.

Stanley, D.J. (1988), Subsidence in the northeastern Nile Delta: rapid rates, possible causes, and consequences. Science 240, 497-500.

Thyagarajam, G. (1983), India Newsletter 46, World Wildlife Fund, Gland, Switzerland.

Tiner, R.W. (1984), Wetlands of the United States: Current Status and Trends. U.S. Fish and Wildlife Service, Washington.

Turner, R.E. (1977), Intertidal vegetation and commercial yields of penaeid shrimps. Tans. Am. Fish. Soc. 106, 411-16.

Turner, R.E., R.L. Costanza and W. Scaife (1982), Canals and land loss in coastal Louisiana. In D.F. Bosch (ed.), Proceedings of the Conference on coastal erosion and wetland modification in Louisiana: causes, consequences and options. U.S. Fish and Wildlife Service, FWS/OBS-82/59, 73-84.

U.S. Corps of Engineers (1972), Cited by J.M. Sather and R.D. Smith, in: An Overview of Major Wetlands Functions and Values. Report for U.S. Fish and Wildlife Service, FWS/OBS - 84/18, September 1984, 68 pp.

van Eck, H., Govers, A., Lemaire, A. and Schaminee, J. (1984), Irish bogs: a case for planning. Nijmegen, Holland, Catholic University.

Wall Street Journal (1990), Florida Set Out to Restore Wetlandsby Refilling and Canal Inadvisably Dug. p. A4, 5 July 1990.

9.

UNDERSTANDING THE FOREST ECOSYSTEM
- A PREREQUISITE FOR SURVIVAL

FOLKE O. ANDERSSON
Dept. of Ecology and Environmental Research
Swedish University of Agricultural Sciences
P O Box 7072, S-750 07 UPPSALA, Sweden

ABSTRACT

Some basic principles of a forest ecosystem are given. From these, changes in the forests are discussed both in a geological time perspective and, in a more recent perspective, as a consequence of the present forest management and air pollution situation.

In particular, recent soil acidification and its consequence for the nutrition of trees and forest growth is discussed. Ongoing changes of flora and fauna in Southern Swedish forests are also discussed.

The need to be able to predict future changes is mentioned. In order to maintain a sustainable future for the forests, regulations must be introduced. For pollution, the concepts of critical levels and critical loads are presented. The future utilization of the forests as a natural resource in industrialized as well as developing countries will require broader planning than ever before by taking into account the different values of the forest ecosystem.

Key Words: *Air Pollution, Flora, Forest Ecosystem, Forest Growth, Forest Nutrition, Future Changes, Historical and Present Changes, Resource Planning, Soil Acidification.*

"Forests were the habitat of our earliest evolutionary ancestors and have remained an important part of the environment of most branches of the human family tree. The rise and fall of empires, the conquest of nations, and the political, economic, and military power of human societies have been intimately related to the accessibility of forests and/or forest products for most of our recent history. While modern Man is no longer truly a forest animal, humans still are, and probably always will remain, dependent on forests for a wide variety of the necessities of life".

J.P. Kimmins. From an introduction to a textbook in Forest Ecology (1987).

INTRODUCTION

The title of this paper may at first glance look overdramatized but, considering historical facts and the way we currently utilize forests in different parts of the world, there is good reason for it. The introductory quotation summarizes the importance of forests over the millennia very well.

In the early stages of development of the human society, the forest was simply a part of the environment: somewhere to live and to find food, energy, and shelter. Simultaneously, the forest was also the habitat of prey and enemies. An increasing population and climatic changes, gradually created the need to exploit the forest when a shortage of wood began occurring. Therefore, wood sources had to be found in other remote areas. Attempts to regulate the exploitation of forests were also made. This led to the birth of forestry, which can be described as non-ecological dictates based upon trial-and-error experience. Gradually, during this Century, an ecologically based resource conservation and management forestry was created. We still do not understand the functioning of the forest in a way which makes it possible for us to give fully reliable advice for managing a forest or to predict its future changes in a satisfactory way.

The aim of this paper is to indicate the present understanding of forests with examples from the boreal and temperate regions. Man's perturbation of the forests will be briefly discussed in a historical perspective which will shed some light on the present situation which will also be analyzed. Finally, the question as to how to predict the future behaviour of forests will be discussed as will the need to wisely plan the use of forests in order to achieve a sustainable future.

The facts used will mainly refer to Scandinavian conditions. However, they are, in many cases, relevant for the Central European conditions as well. Relevant references are also given to papers documenting conditions outside Scandinavia.

THE FOREST ECOSYSTEM - PRINCIPLES

An integral part of understanding ecosystems is to discover how the processes responsible for the turn-over of energy and matter are regulated and how an ecosystem develops in relation to time and space. The ecosystem is characterized by a high degree of complexity with interactions and interdependencies between all the organisms and the non-living environment. These interactions, between living and non-living parts, make it difficult to predict the behaviour of a process or of the ecosystem itself.

The concept of the ecosystem originates from the 1930's (Tansley, 1935) and the principles of ecosystem ecology were brought forward during the 1950's (Odum, 1953, 1964). Agriculture and forestry may be regarded as applied ecology. It was not, however, until the last two decades that a more functional and mechanistic research on an ecosystem was introduced. Nevertheless, a considerable wealth of knowledge exists on the different parts of the forest, such as: the soil, the tree, its physiology and genetics, forest regeneration, forest production, etc. However, the mechanistic understanding of forest growth is still incomplete.

FUNCTIONAL ASPECTS OF A FOREST

In a wider sense and from a human utilization point of view, we may list a number of ecosystem functions by the following short characterization:

Production function: The ability of the ecosystem to form organic material from inorganic compounds. Here we consider plant and animal production.

Carrying function: Spatial utilization of ecosystems or different forms of land use.

Storage function: The ability of ecosystems to store water, organic and inorganic materials. The energy storage in organic material is essential.

Recipient function: The ability of ecosystems to receive and break-down waste and surplus products, often in the form of pollutants such as gases, particles, radiation and heat.

Protection function: A number of different protection functions from cosmic radiation to military function.

Information function: Here we consider different functions ranging from gene banks to the stimulus for human activities. The information is mainly coupled to the biotic component of the ecosystem.

Regulatory function: Includes the reaction of the ecosystems to changes, which may be buffered and thus contribute to the balance and stability of ecosystems. The information and the regulatory functions are the most important ecological functions of the environment.

ENVIRONMENTAL CHANGES AND MAN

Changes in a Geological Time Scale

In our daily life as inhabitants of this planet, we often assume that environmental conditions are stable. We know, of course, that the weather is changeable from year to year and that drought situations can occur in intervals of a decade or decades. It is, however, more difficult for us to follow and perhaps also accept more long-term changes. Therefore, a few examples of changes occuring on the Globe and discussions on how Man may interfere, in particular, with forests at our latitudes will be given.

The first example will show the content of carbon dioxide in the atmosphere during the last 160,000 years (Figure 1). It is regulated by the diffusion of carbon dioxide in the oceans and the photosynthesis and respiration of plants. The data was derived from cores taken in the inland ice of the Antarctic and Greenland. The three phases can be identified as follows:

- 150,000 years ago the levels of carbon dioxide were very high for reasons not known. A period with levels changing between 200 and 270 ppm then followed

- Since the last glaciation, 20,000 years ago up to the beginning of the 1800's, a stable level of carbon dioxide, 260-290 ppm, has persisted

- Since 1800, the carbon dioxide content has steadily increased up to its present level of 350 ppm. Simultaneously, other heat-absorbing gases, such as freons, methane, dinitrogen oxide and ozone, have increased.

This last increase can be attributed to Man and his activities due to the changing land use and the combustion of fossil fuels. The environmental discussion nowadays deals with the consequences of this increase.

Post-Glacial Changes

In order to understand the present situation better, a historical perspective may be rewarding. A number of glaciations have followed each other. Let us examine what happened after the last glaciation here in Europe. Ice covered Northern Europe down to the Northern part of Germany, Luneburger Heide and Poland, and the Masurien Lakes. The Northern parts of the British Isles were also glaciated as were the Ural Mountains.

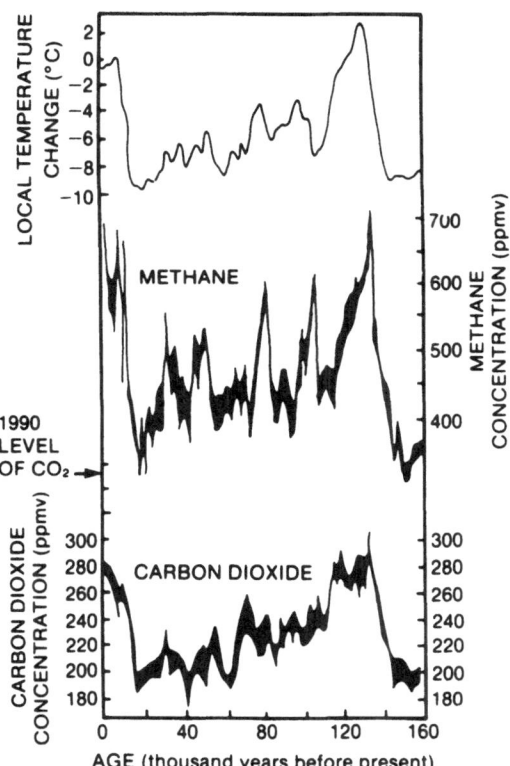

Figure 1. Carbon dioxide and methane concentrations in air trapped in the Antarctic ice. A correlation with local temperature is demonstrated during 160,000 years.
Present day carbon dioxide level is indicated. From: Anonymous, 1990a.

We are interested in how forests and soils have developed after that event. From pollen analyses and the emmergence of different plant species, it has been possible to determine how the climate developed (Figure 2). Arctic periods were followed by warmer conditions that were either more moist or more dry. Later on cooler and wetter periods prevailed.

As time elapsed, the ground, which not became covered by forest canopy, once again was made more open, due to the arrival of Man. The soils became more acidic as basic elements like calcium and magnesium were consumed by the vegetation. Furthermore, the decomposition of litter also resulted in a formation of acids acidifying the upper part of the soil.

The first arctic, non-leached soils gradually started to change (up to preboreal time - 7700 BC). During the boreal, Atlantic and sub-boreal periods (7700 - 600 BC)

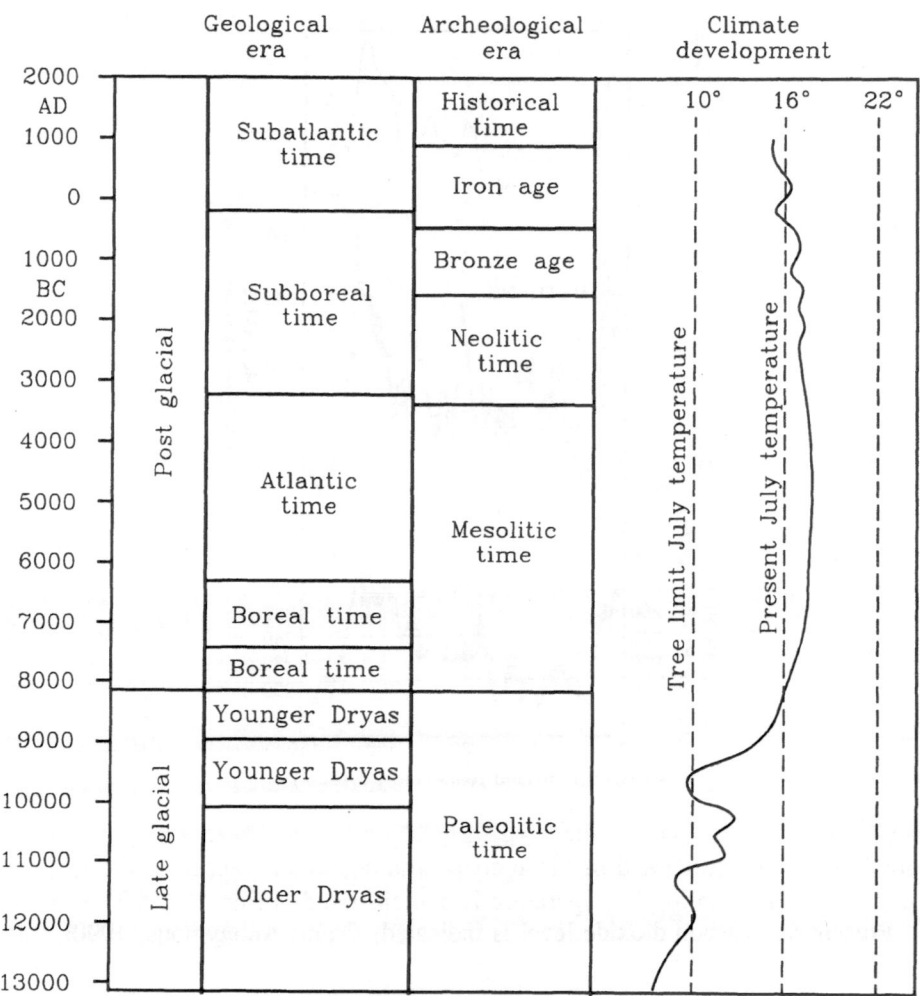

Figure 2. Chronologic survey of Late Quaternary time in Denmark and Skåne, Southern Sweden. From: Berglund, 1968.

brown-forest soils or cambisols dominated. These soils were weakly acidic and are the soils of the richer deciduous forests. With a decreasing temperature and increasing humidity - the subatlantic time or the present time - the soils continued to change. The soils generally became more acidic and podsols (spodosols) - the soil type of the boreal coniferous forest - became dominating (Figure 3).

Man's utilization of the forests or the landscape is usually described in a number of phases, which are called "landnam", using the Danish term. Man's exploitation of the Southern Scandinavian forest landscape is usually described in six phases as follows:

1. Early neolithic time (3300-2500 BC). Only local effects are observed. In a pollen diagram this is seen as a decrease in the occurrence of elm.

2. Up to the early Bronze Age (2200-1000 BC) the keeping of cattle and farming increases. Formation of heathlands.

3. During the early Iron Age (200 BC-450 AD) a continued expansion of open land. Signs of overexploitation.

4. The present landscape is formed during the late Iron Age (800-1100 AD). A drastic climate change occured - the Fimbul winter. The need for winter fodder increased. Expansion of mowing and grazing land. The formation of villages with in- and out-fields.

5. Apart from temporary declines in the activities of Man, the landscape became more open due to the need of more arable and grazing land. Big projects to create ditches and lower water levels in lakes in order to obtain more arable land are undertaken. The landscape reaches its maximum openness during the middle of the 19th Century.

6. Agricultural methods become more effective due to the continued making of ditches, drainage of land and, in particular, the use of artificial fertilizers. The area of arable land decreases, forest land increases.

Man's presence in the landscape can be followed since 3000 BC. The initial stages 1-4 illustrate fairly moderate impacts (Figure 4). Later, increasing grazing and shifting agriculture with burning has a considerable impact and creates an impoverished landscape.

We are now entering a new era as a result of the abandonment of arable land. The forest area will increase again. The forest soils in Southern Scandinavia are changing at a greater speed than earlier as a consequence of the air pollution, as are the flora.

Present Changes
Air Pollution. Atmospheric pollution was, in the beginning of the industrialization, originally of local origin - a point source phenomenon. Changing technology caused higher chimneys to be built and mobile combustion sources to be invented - the cars. The original local phenomenon thus then became a regional one. Today it is calculated that 10% of the terrestrial part of the globe is impacted. It would be more correct though to say that air pollution is a global phenomenon because the emission of carbon dioxide and other trace gases are affecting the heat balance of the entire globe.

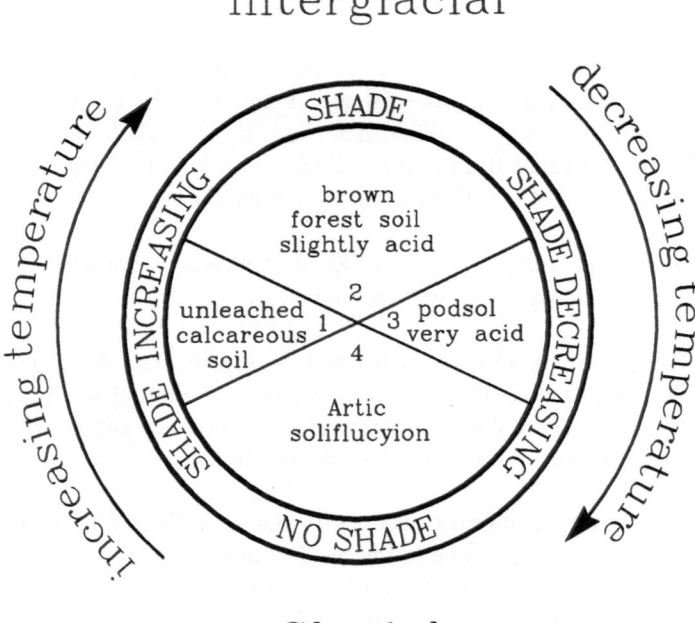

Figure 3. The glacial-interglacial cycle according to Iversen, 1958, indicating general temperature, light and soil conditions.

We will take a closer look at the effects of the major industrial pollutants, i.e. sulphur and nitrogen. Firstly an example of the emissions of sulphur and nitrogen will be given for West Germany (Figure 5). Emissions are expressed as the conversion products of the original sulphur dioxide and nitrogen oxide, sulphuric acid and nitric acid, respectively (unit: kg-equivalents hydrogen/ha and year). The diagram illustrates the industrial activities from 1850 until 1980. Decreased activities are seen as a result of World War I and II as well as the economic recession of the 1930's. Note the dramatic increase of pollutants and culmination of pollutants during the post-war period. The figure also includes the estimated, natural background levels together with the release of buffering elements in the soil. The information is easily interpreted to mean that the emissions will unavoidably effect our terrestrial environments.

A number of characteristic deposition regimes can be identified (Table 1). They have been identified in such a way that the areas with high concentrations of sulphur dioxide and ozone with risk for direct effects on organisms are easily seen. The levels of dry and wet deposition of sulphur and nitrogen are then of interest as they indicate where predominantly indirect effects occur. The indirect effects are operating through the soil.

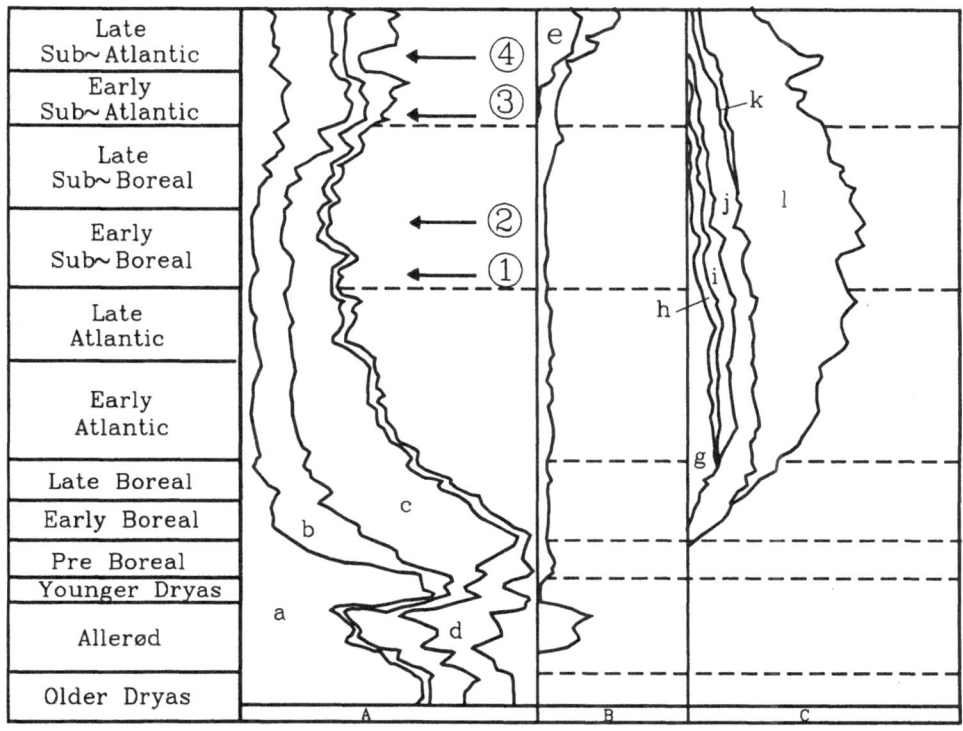

Figure 4. Simplified pollen diagram showing main features in the vegetational development of the coast area in Eastern Blekinge, Southern Sweden. A: Plants which do ot tolerate competition and shade; B: Plants indicating acid humus; C: Plants indicating better humus (mull). Arrows 1-4 indicate times of more intensive human influence. a: Herbs and graminids; b: *Betula* (excl. *B. nana*); c: *Pinus*; d: *Populus* + *Juniperus* + *Salix* + *Ericales*; e: *Fagus*; f: *Ericales* + *Pteridium* + *Sphagnum*; g: *Ulmus*; h: *Fraxinus*; i: *Tilia*; j: *Corylus*; k: *Carpinus*; l: *Quercus*. From: Berglund, 1969.

The regional, mainly indirect, effects on forests have been and still are discussed. In many cases they are difficult to measure and identify, because the forest ecosystem reacts to all factors. The health of the forest or the growth of the forest is a result of the combined effects on the system, climate, pollution and soil. With an increased exposure time, the effects of pollution have become more obvious. We will look more closely at some Southern Swedish information, which illustrates how forests indirectly are affected by pollution. Even toxic effects may occur when some heavy metals, such as aluminium, gets released at low pH-levels.

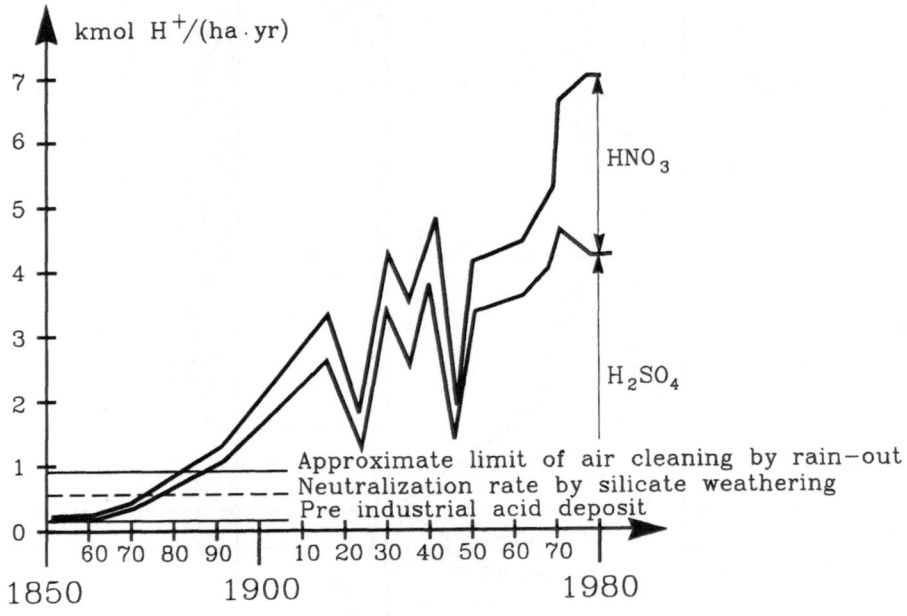

Figure 5. Emissions of sulphur (sulphuric acid) and nitrogen (nitric acid) in the area
    of the Federal Republic of Germany since 1850. From: Ulrich 1989.

Soil Acidification
We have already seen that in the soil development after the last glaciation, acidification
appeared as a natural phenomenon. Acidification is normally caused by biomass production (uptake of basic cations leading to a decrease in the soil) and when organic
matter is decomposed. Changes leading to a more humid climate have provided an
environment which favours acidification due to the leaching of elements. It is also
natural that a young forest has a higher pH in the upper parts of the soil compared to
an old forest.
      In a unique study on the Swedish west coast it has been possible after 60 years,
to repeat a study of soil conditions in 1927 (Tamm and Hallbäcken, 1988). It has also
been possible to compare this study with a similar one from Northern Sweden, thus
providing a comparison of soils under different pollution regimes. In the Southwestern
Swedish case, drastic changes of soil pH in both upper and lower parts of the soil in
forests of beech and Norwegian spruce were found. The changes varied from 0.7 - 1.5
pH-unit (Figure 6). The natural or biological acidification due to the increasing age of
the forest is also demonstrated (Figure 7). We can observe that all pH-data for the
1980's are lower than the original ones.

Table 1
Characterization of deposition situations of Central and Northern Europe. Information compiled from different sources by the Author.

Region	Total S-dep. kg/ha	SO$_2$ $\mu g/m^3$ yr.- mean	pH	O$_3$ hr > 150 $\mu g/m^3$	Total N- dep. kg/ha
1. Western CSSR, Ruhr, Western Belgium, Central England	>60- 120	>20	<4.1	100-200	10-20
2. Western & Central Europe other than area 4 and 5	30-60	15-25	<4.1	100-200	20-60
3. Eastern and Central Europe	30-60	15- >25	4.1-4.5	100-200	20-40
4. Southern Germany - lowland	20-40	<10	4.0-4.5	100-200	10-20
5. Southern Germany high altitudes	35-50	<10	<4.1	>325	10-20
6. Southern Scandinavia, Southern Finland, Baltic States, parts of UK	15-30	8-15	<4.1 4.1-4.5	50-100	20-30 10-20
7. Central Scandinavia and Central Finland	8-15	<8	4.3-4.6	10-50	5-15
8. Northern Scandinavia and coastal areas	<8	<3	>4.6	?	<5

In the Northern Swedish case, no pH change can be seen for the deeper part of the soil as is the case in normal soil development. The Southwestern Swedish forests receive a higher acidity of acidifying sulphur and nitrogen compounds. In order to maintain electrical neutrality, mobile sulphate and nitrate anions (negatively charged) take positively charged cations from the soil colloids when the water percolates through the soil. Hydrogen and aluminium ions are taken up instead. A regional soil acidification is the result (Berdén et al., 1987). Experimental results (Tamm and Popovic, 1990) support the explanation that it is the anthropogenically produced air pollutants which are responsible for a large-scale soil deterioration (Berdén et al., 1987). The air pollution increases the loss rate of the elements, which corresponds to an increased degree of podsolization.

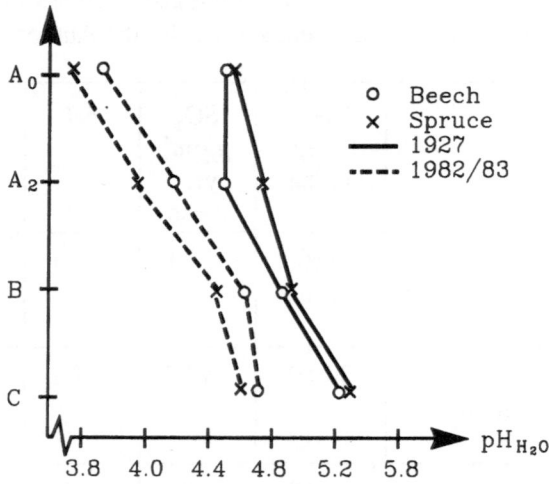

Figure 6. Average pH changes during a 50 year period in soils of beech and Norwegian spruce forests at the Tönnersjöheden Experimental Forest, Halland, Southern Sweden. A_0 and A_2 = humus horizon, B = accumulation horizon in mineral soil, C = original, non-affected mineral soil. Generalized from Hallbäcken and Tamm, 1985.

Available Nutrients and Tree Growth

From other investigations in Southern Sweden it is possible to conclude that there is an unexpected loss of nutrients from the soil due to the soil acidification (Falkengren--Grerup, 1990). Up to 50% of the mineral nutrients (calcium, magnesium, etc.) usually available to the plants have been lost (Figure 8).

The long-term consequences of this, if the soil does not have the ability to make more minerals available to plants by weathering, will be a deficiency situation. Most likely deficiency symptoms on leaves and decreased growth will then appear due to a lack of mineral elements such as phosphorus, potassium, and magnesium.

Indications of changes in the nutritional situation of trees have been demonstrated in Southern Sweden over a period of 20-30 years (Aronsson, 1985). Nitrogen is increasing as a consequence of an increased deposition during the last 40 years. Phosphorus, calcium, and zinc are decreasing as a consequence of changes in the soil.

In Scandinavia, so far, no general decrease in forest growth has been found (Tamm, 1989; Andersson, 1989). A compensatory effect of the increased nitrogen deposition may be a reasonable explanation of this. However, the long-term productivity of our forests is endangered. Soil acidification leads to disturbed conditions for nutrient uptake and lower amount of available nutrients. Rosén (1989) has compared nutrient budgets for forests in Southern and Central Sweden, areas with high and low deposition of air pollution respectively. The annual loss of essential nutrients such as calcium is 2-3% per year. Today we cannot see symptoms of nutrient deficiencies, however the next forest rotation with its approaching canopy closure and high nutrient demand may be subject to severe nutrient stress which may also lead to decreased long-term productivity.

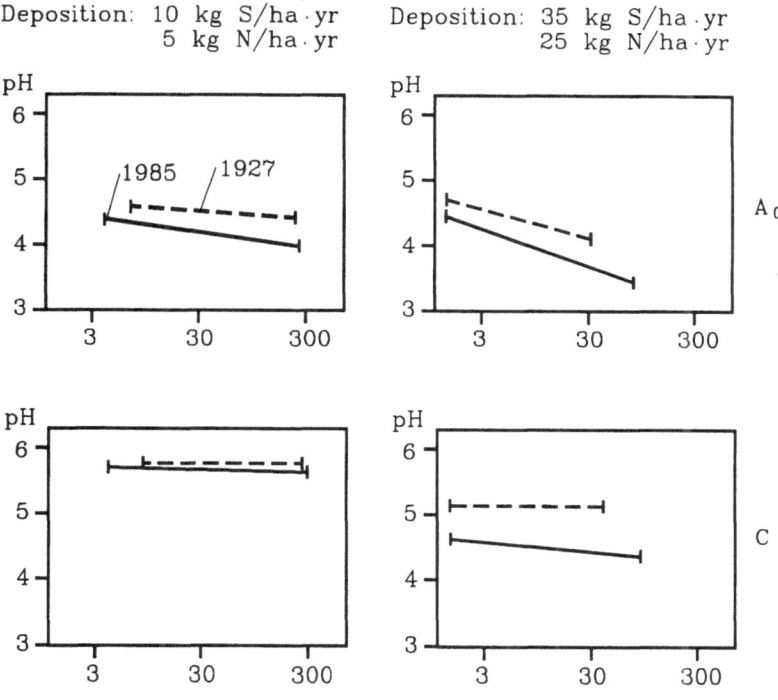

Figure 7. Variation of pH with logarithmic stand age (years) in A_0-humus (upper part) and C horizon (lower part) at Kulbäcksliden, Northern Sweden, and Tönnersjöheden, Southwestern Sweden. Generalized from: Tamm and Hallbäcken, 1988.

Changes in Flora and Fauna.

In the same way that the understanding of ongoing soil changes has increased, we are, at present, experiencing a time where the evidence is accumulating that flora and vegetation are changing as a consequence of air pollution (Andersson, 1989; Torstensson and Liljelund, 1989). Descriptive investigations show that epiphytic lichens and mosses are decreasing or have disappeared (Löfgren and Moberg, 1984; Hallingbäck, 1986; Olsson and Hallingbäck, 1987). Demanding species decrease with an increasing soil acidification. Nitrogen favoured species increase with increasing deposition. It is also assumed that the fungal flora is changing in a similar way. In the individual case it is difficult to tie a change of a species to a certain cause. However, the accumulated evidence suggests that air pollution is either directly or indirectly responsible for the observed changes (Falkengren-Grerup, 1990).

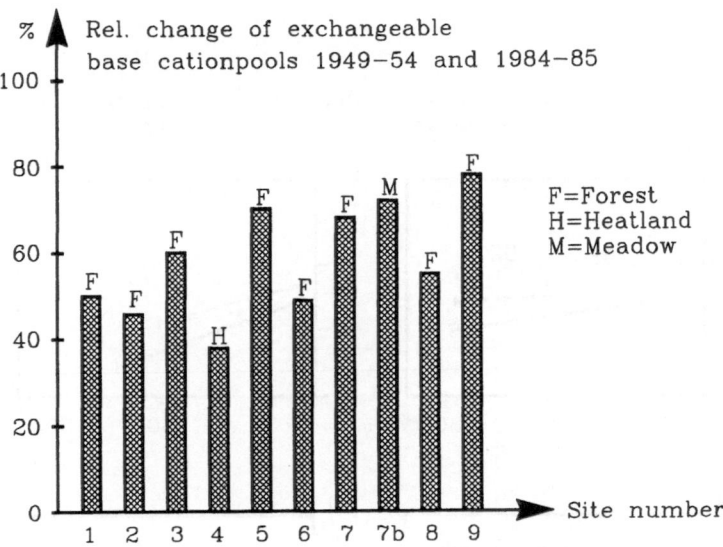

Figure 8. Relative change of exchangeable base cation pools 1949-1954 and 1984-85. The sites are ordered from the most acid to/ the least acid soil. Redrawn from Falkengren-Grerup et al., 1987.

Changes in the fauna are also reported but to a lesser extent. Amphibians decrease as a consequence of water acidification Molluscs are also decreasing in number and abundance.

It is logical to assume that the ongoing soil acidification and the present deposition of nitrogen will lead to the following general pattern of change: rare, demanding species will decrease and less demanding and nitrogen-favoured species will increase. The result will ultimately be an impoverishment of both flora and fauna.

At present, in Sweden, there is a grave concern about the future flora and fauna of the forest, not only as a consequence of air pollution, because the present forest management seems to lead to a less floristically varied landscape.

FUTURE CHANGES

The presentation so far has indicated that over the millenia there have been natural changes of our environment but that, during recent centuries, Man has contributed to these changes. The present utilization of the forest leads, without a doubt, to changes. There is a need to understand the consequences of different intensities of management in order to maintain not only the biological diversity but also the long-term productivity of the forests. In addition, the impact on climate and ecosystems through combustion of organic material and fossil fuels will most likely lead to unexpected changes of the forests.

What will the future consequences be? For this part of the industrialized and temperate part of the world, the forests are presently changing as a consequence of past use and present impact of pollution. The coming climatic change will make the predictions of the future still more difficult.

Will the forests maintain their productivity? A forester usually uses his experience and his yield tables to tell him how much a given site will produce. However, these tables are based on historic data collected under different environmental conditions. Today the air pollution, if it does not kill the forests, mainly changes the nutritional conditions. In many parts of Europe, Northern Europe, we experience a better forest growth as a consequence of the fertilizing effect of nitrogen (nitrate and ammonium). Growth increase may also be stimulated by the increased carbon dioxide level. Sooner or later, however, other nutrients will eventually be limiting, and a decreased forest growth will be the result.

INTRODUCTION OF CRITICAL LEVEL AND CRITICAL LOAD CONCEPT

The present understanding of trees and soil in forests has made it possible to introduce critical levels and loads of pollutants, which ought not to be exceeded if sustainable conditions for organisms and productivity should be maintained (Nilsson and Grennfelt, 1988). These concepts are important for international negotiations as well as for national goals of establishing environmental aims or targets.

The following international standards, annual means, have been agreed upon if trees are not to be injured by gases such as sulphur dioxide and ozone:

sulphur dioxide $< 25 \ \mu g \ m^{-3} \ y^{-1}$ (μg per m^3 and year)
ozone $< 50 \ \mu g \ m^{-3} \ y^{-1}$

The critical load concept describes the acceptable level which can be deposited on terrestrial ecosystems. For sulphur, if lakes in sensitive areas are not to change further, a critical load of app. 5 kg S ha^{-1} y^{-1} (kg S per hectare and year) can be accepted. As to the soils, a further soil acidification will also be avoided. For Southern Swedish conditions the figure should be compared to the present 20 - 35 kg S ha^{-1} y^{-1}, a further reduction of 4 - 7 times is thus required. In Northern Sweden the present deposition is almost acceptable for the environment.

For nitrogen, high deposition levels may lead to a situation where the forest cannot take up all nitrogen. We then get a leaching of nitrate into streams, lakes and the sea. Acceptable levels, depending on site conditions, amount to 10 - 20 kg N ha^{-1} y^{-1} (kg N per hectare and year). Present figures in Southern Sweden are 15 - 25 kg N ha^{-1} y^{-1}. The present deposition in Northern Sweden is within acceptable levels.

This quantification of acceptable levels for the impact of pollutants on forest ecosystems has been a major achievement which provides a strong background for a discussion on measures to maintain a healthy and sustainable environment for all organisms, including Man. In time, these concepts will be improved.

NEED FOR NATURE RESOURCE PLANNING

From time to time, an intensive debate occurs between environmentalists and representatives of forestry. In Sweden, such a debate occured during the 1970's and now again occurring in 1990. The major concern is that modern forestry practices are a threat to the biological diversity in the forest and also a threat to the long-term productivity.

Sweden is a forest country. We have a forestry law, which favours forest production and within this law special environmental consideration must be shown with regard to landscape, fauna and flora. In situations where we have agreed to international conventions (WCS - World Strategy of Nature Conservation 1980, Bruntland report 1988, Environmental program of the Nordic countries 1989) it is now necessary to develop a strategy where, as a base, we have a natural resources strategy. Within this framework the forestry must operate (Anonymous, 1990b).

The major points in such a nature resource strategy are:

-    maintainance and in some cases, redevelopment of genetic and ecological diversity and variecy, i.e. preservation of species, biotopes and forest types in the future forest landscape.

-    well-functioning cycles of energy, water and nutrients within and between soil, vegetation, fauna, water and atmosphere.

-    long-term and sustainable utilization of soil, forest and water in accordance with a sustainable development.

-    a multiple use of the forests.

If we now wish to practice a forestry within an expressed nature resource strategy there is a strong need for a new plan to utilize the forest landscape. Traditionally in Sweden there are good planning instruments available for the utilization of the forest for production of timber and fiber. What is now required is a broader aspect. We need to know the requirements of flora and fauna, consider the consequences of various forest management practices, not only on the flora and fauna but also on water, nutrients and landscape values. Within this framework, the threat from pollution, climate change and other forms of utilization of the forests must be considered.

Society needs to set the aims and the criteria for the future utilization. The scientists shall provide the necessary background for these criteria and also develop the adequate planning tools.

CONCLUDING REMARKS

The use and value of the forests can be different in different parts of the world. It can be a question of finding fuel wood for the daily need and existence as in some developing countries. It can be a question of deriving consumption and export goods to maintain a high GNP and standard of living as in the home country of the Author. It is for a future sustainabilty of production and biological diversity important that the use of the forest is handled in a wise way. Therefore, an element of forestry as well as biological planning is needed.

We must avoid creating unnecessary stress on the forests. A utilization of the forests in and of itself results in itself in different forms of stresses, which cannot be avoided. Unnecessary stresses such as air pollution can, however, be avoided.

A control of the pollution to acceptable levels is possible and requires solidarity between nations. The utilization of the forests must be in harmony with the conditions of the region. Therefore, it is not advisable to introduce foreign elements which may lead to impoverished conditions of biological diversity or productivity.

REFERENCES

Andersson, F. (1989a), Swedish forests in a changing environment - effects on soils, tree nutrition, and growth. In: Forest health and productivity. The Marcus Wallenberg Foundation Symposia Proceedings 5: 29-49, Falun. ISSN 0282-4647.

Andersson, F. (1989b), Air pollution impact on Swedish forests - present evidence and future development. Environmental Monitoring and Assessment, 12: 29-38.

Anonymous (1990a), Scientific assessment of climate change. The policy makers' summary of the Working group I to the WMO/UNEP Intergovernmental Panel on Climate Change, Geneva.

Anonymous (1990b), Reinforcement of research of the area forest production-environment-nature. - A report prepared by a task force appointed the Forest Faculty. - Uppsala, Swedish University of Agricultural Sciences. (In Swedish).

Berdén, M., Nilsson, S.I., Rosèn, K., and Tyler, G. (1987), Soil acidification -extent, causes, and consequences. National Swedish Environment Protection Board Report 3292, Solna.

Berglund, B.E. (1968), Vegetationsutveckling i Norden efter istiden (Vegetation development in Nordic countries after the last ice age). Sveriges Natur Årsbok, pp.31-52, Stockholm.

Berglund, B.E. (1969), Vegetation and human influence in South Scandinavia during prehistoric time. Oikos Suppl., 12: 9-28.

Falkengren-Grerup, U. (1990), Soil acidification and vegetation changes in South Swedish forests. Doctoral dissertation. 18pp. LUNDDS(NBBE-1033)/1-18/(1989).
Hallingbäck, T. (1986), Lunglavarna, Lobaria, på reträtt i Sverige. Svensk Bot.Tidskr., 80: 373-381.

Hallingbäck, T. and Olsson, C. (1987), Lunglavens tillbakagång i Skåne. Svensk Bot.Tidskr., 81: 103-108.

Iversen, J. (1958), The bearing of glacial and interglacial epochs on the formation and extinction of plant taxa. Systematics of today. Proc. Sympos. Univ. Årskr. 1958, 6: 210-215, Uppsala.

Kimmins, J.P. (1987), Forest ecology.

Löfgren, O. and Moberg, R. (1984), Oceaniska lavar i Sverige och deras tillbakagång (Oceanic lichens in Sweden and their retreat). Swedish Environment Protection Board Report 1819, Solna.

Nilsson, J. and Grennfelt, P. (1988), Ed: Critical loads of sulphur and nitrogen. Report from a workshop held at Skokloster, Sweden 19-24 March, 1988. Nordic Council of Ministers. NORD 1988: 15. ISBN 87-7303-248-4.

Odum, E.P. (1953), Fundamentals of ecology. 1st ed. Philadelphia: Saunders (2nd ed. 1959, 3rd ed. 1971).

Odum, E.P. (1964), The new ecology. BioScience, 14: 14-16.

Rosén, K. (1989), Influence of sulphur and nitrogen deposition on base cation supply in managed forests. In: W.J. Dyck and C.A. Mees (Ed.). Research strategies for long-term site productivity. Proceedings, IEA/BE A3 workshop, Seattle, WA, August, (1988. IEA/BE A3 Report No.8. Forest Research Institute, New Zealand, Bulletin 152, pp. 165-172.

Tamm, C.O. (1989), Comparative and experimental approaches to the study of acid deposition effects on soils as substrate for forest growth. Ambio, 18: 184-191.

Tamm, C.O. and Popovic, B. (1990), Acidification experiments of pine forests. National Swedish Protection Board Report 3589, Solna.

Tamm, C.O. and Hallbäcken, L. (1988), Changes in soil acidity in two forest areas with different acid deposition; 1920's and 1980's. Ambio, 17: 56-61.

Tansley, G. (1935), The use and abuse of vegetational concept and terms. Ecology, 16: 284-307.

Torstensson, P. and Liljelund, L-E. (1989), Flora- och faunaförändringar i terrsetra miljöer orsakade av luftföroreningar och försurning. National Swedish Environment Protection Board Report 3604, Solna.

Ulrich, B. (1989), Effects of acid deposition on forest ecosystems. In: Forest health and productivity. The Marcus Wallenberg Foundation Symposia Proceedings, 5: 5-28, Falun.

10.

THE CONSERVATION OF BIOLOGICAL DIVERSITY: USING BIRDS TO SET PRIORITIES

JON FJELDSÅ
Zoological Museum, University of Copenhagen
Universitetsparken 15, DK-2100 Copenhagen, Denmark

ABSTRACT

Biological diversity means the diversity of living beings. Especially in the tropics, the immense variation of species is now rapidly being reduced by man. This may lower the resilience of ecosystems and means irreversible loss of potential resources. The crisis has now reached such daunting dimensions that the traditional focus on single species must be abandoned. Instead, the resources should be used for securing ecosystems and pooling the effort into areas that are identified as having key importance for maintenance of biodiversity. The approach is facilitated by the fact that many local (endemic) and non-dispersive species cluster together in certain areas in the tropics. Assuming that centres of endemism signal where new species evolved by isolation of once-widespread forms it is postulated that centres of endemism in the tropics represent places with maximum capacity for maintaining stable ecological regimes over sustained time periods. These may be optimal sites for maintaining life support systems in a wider sense.

Only birds have been studied and charted with sufficient uniformity for a global survey of patterns of endemism. A project compiling and analyzing the current data on endemic birds, supplemented with surveys of unexplored portions of potential priority areas and consensus of judgement from other life-form experts, may become a central tool for global planning. Charting of other selected groups of organisms will certainly reveal other or finer patterns, but for feasibility reasons should be done primarily in areas which seem valuable from indications already available.

Key Words: Biodiversity, Biological Conservation, Endemism, Ornithology.

INTRODUCTION

The term biological diversity (= biodiversity) covers the sum of different living organisms - animals, plants and microorganisms. The collective stock of species, worldwide, is immense. Recent estimates show 30 to 50 millions of kinds (Erwin, 1983, May, 1989). Ironically, the term became common usage as a result of the deepening discomfort over the man-made acceleration of the rate of extinction of species. It is estimated that 25 to 50 percent of all species will disappear during the next hundred years (Ehrlich, P. and A., 1983, Myers, 1983, 1984). Biologists see this result of "development" as one of the great crises of our planet, and even politicians, the World Bank, etc. now ask whether we can afford the risk of loosing great parts of the existing species (McNeely et al., 1990, U.S.Congress O.T.A., 1987, World Commission on Environment and Development, 1987).

First of all, there are principal, ethical and aesthetic arguments for saving species: the value of living beings per se (see Wilson, 1984). But considerable human self-interest exists as well (e.g., Ehrlich, P. & A., 1983, Myers, 1983, 1984, Fitter, 1986, Prescott-Allen, 1988, Wilson, 1985). Species have been seen as parts of a self-regulating whole. The diversity is partly responsible for the resilience of ecosystems. Species also represent resources: A large proportion of our most important medicines are based on chemicals that plants use for self-defence. A dozen plant species which happened to be chosen for cultivation by early human cultures still supply 90% of the world's food, and just three - wheat, maize and rice - constitute about half. Yet, only a tiny fraction of living organisms has yet been tested for practical application. Gene technology has greatly expanded the basis for a material use of biodiversity to enhance the resilience of cultivated forms.

Since the crisis and its consequences is well described in numerous publications, I will not enlarge on this, but rather on the possible cures. Their efficiency will depend on how rapidly we can identify the top priorities and plan cost-effective actions. The present paper is a personal view of a model for how to set priorities.

(Indeed, environmental conservation has numerous facets, but I will limit myself here to the question of how to reduce the most irreparable damage: the total extinction of species. I will discuss terrestrial ecosystems only. Strong emphasis will be placed on birds. This is not because I regard birds as a particularly valuable part of the biodiversity, but because the world's birds have been recorded sufficiently completely to make bird data a suitable basis for immediate identification of top priority areas.)

APPROACHES TO COMBAT THE BIODIVERSITY CRISIS
- FOCUS ON SINGLE SPECIES

The traditional approach to halt the biodiversity crisis included actions for special or big, charismatic species. The single-species approach continues to attract attention (see, e.g., Soulé, 1987). "Sexy" species are useful for fundraising and as "flagship species" for causing awareness of the threats to the habitat. However, many single-species projects are nothing but symptoms treatment. In these cases, they do not solve the fundamental problems. If the assertion that something is done distracts attention from the real issue, the projects become counterproductive.

The perspectives of maintaining artificial collections in zoos and botanical gardens and storing seeds or germ plasm are a last resort for preserving a selection of already known forms. Successful re-introduction into the nature of captive-bred individuals seems, in general, to be much more difficult than the captive-breeding as such. It works well only as part of a broad, integrated approach, and therefore seems to represent a realistic solution only for a tiny fraction of the existing species (Imboden, 1987). Finally, the number of species in need of conservation action has become so daunting that it is hopeless to make specific actions for them all. It must therefore be considered whether such projects really represent wise use of the money available for conservation.

FOCUS ON AREAS RICH IN SPECIES
OR WITH MANY ENDEMIC SPECIES

As we still only know small parts of the biodiversity, an obviously more efficient approach would be to identify areas that seem to have special importance for the evolution or maintenance of biodiversity, and then to secure the continued ecological functioning of these areas. This choice could be made by a coordination of efforts by various experts, but primarily students of plants or animals that are suitable for rapid surveys.

In the past, the selection of areas for conservation was often dictated by conflicts with economical interests, or by prospects for tourism. Many protected areas fully qualify for titles such as "World Heritage Sites", but are not necessarily good choices for the sake of reducing the rate of extinction. With constant conflicts with development interests, decisions must be made as to which parts of the world are best used to feed mankind, and which are to be set aside to secure ecological processes and biodiversity. It is obviously necessary to examine with explicitness where a maximum number of species can be "saved" on a minimum of areas, using the limited funds that are available.

Fortunately for this decision the richness of species is not at all evenly distributed. Vast parts of the earth are inhabited by widespread and adaptable species which may rapidly recover, if tolerable conditions are found locally. This is the general situation in much of the northern temperate zones. For this reason, the almost total destruction of virgin nature in northern Europe has not at all had such irreparable biological consequences as the destruction of virgin tropical forest, with its concentrations of specialized forms of often restricted distribution (endemic species).

We should in the following consider what is the most important: large numbers of species or centres of endemism. For this decision, a brief synopsis of some current views of evolution and maintenance of biodiversity is needed.

THEORETICAL FRAMEWORK

The reductionistic community ecology of the 1960-70s viewed the number of species sharing an area as a compromise between productivity and competitive interaction between species (see e.g. Cody and Diamond, 1975). However, the diversity of species can be raised by temporal and spatial heterogeneity of an area (Connell, 1978, Stenseth, 1980, Salo et al., 1986). Because many species of heterogeneous environments are "programmed" to make opportunistic responses to change (see e.g. Rotenberry and Wiens, 1980), this kind of species-rich system has considerable resilience and most species are widespread. In such ecosystems, the number of species, per se, is not a good measure for the urgency of conservation action. The belief is gaining ground that current ecological processes may be secondary to factors in the past history for determining the number of species (Ricklefs, 1989). It is widely found that differences in species richness between similar ecological zones on different continents are results of different opportunities for generating new species (review by Keast, 1985).

New species normally arise because of chance colonization of remote areas (peripheral speciation) or contractions of habitat, e.g. as long-term climatic changes isolate remnants of once-widespread species (speciation by vicariance). As the habitat expands again, the isolated populations are brought back into contact, sometimes to hybridize (an important mechanism for generating new species of plants!), at other times to confront each other as now separate species. Modern biogeographic analysis uses congruence of superimposed distributions of several endemic species to locate centres of dispersal. Thereafter, they use hypotheses of evolutionary relationships for several groups of related forms to construct "generalized area cladograms" that show historical relationships between areas.

The paradigm for the species-rich tropical forests is that most speciations happened in periods of cool and dry climate (during the ice ages on higher latitudes) when the forests formed relict patches (Haffer, 1974, Prance, 1978, 1982, Diamond and Hamilton, 1980, Prigogine, 1989). Dispersal took place mainly in warm and humid periods. Restriction to a small area, which differs from the regional environment, may so weaken the adaptability and competitive dominance with the regional biota that new species fail to use later opportunities to spread. They remain endemic to a small area.

(Although the "refuge theory" for tropical forest faunas is in general agreement with palaeoclimatological data (Flenley, 1979, Ab'Saber, 1982, Hamilton, 1982, Van Zinderen Bakker, 1982), few hard data exist for locating the past forest refuges. It is also unclear how much of the speciation has underlying unifying causes and how much is a collection of special, unrelated events (Major, 1988). Alternative (or supplementary) mechanisms are: isolation by mountain ranges or major rivers intersecting tropical forest habitats (Lynch, 1988)).

Figure 1 exemplifies the occurrence of endemic and shared species of genuine montane-forest birds of the main mountain tracts of tropical Africa. The endemism is centred in Cameroon, around the Albertine Rift and in the Uzungwa Mountains in the Eastern Arc Mountains. The mountains of Kenya have few endemics and the lush forests of Mts. Meru and Kilimanjaro have no endemic birds at all. Maybe such mountains had, periodically, more open woodland habitat and could be invaded from the surrounding savanna, or colonized by the more adaptable and dispersive of the montane forest birds that used the gallery forests of the savanna zone to move from the Usambaras, Uzungwas or the mountains of Kenya. It could be argued that the mountains with few endemics are volcanoes of few millions years' of age. However, the fact that the majority of endemic birds are members of groups of closely related species that probably differentiated in the course of the last few hundred thousands of years contradicts that the age of the mountain per se is significant. A similar biogeographic pattern as seen in Figure 1 is found for studied groups of insects and spiders.

Since, apparently, species evolve by periodic isolation in certain mountain areas, and thereafter more or less efficiently disperse to others, it is evident that our effort to save many species should be concentrated in those areas that generate species.

WIDENING THE PERSPECTIVE

As a logical deduction from the "refuge paradigm" I believe that particular tracts of tropical forest generate new species because they had the capacity to maintain patches (at least) of relatively unaltered habitat over extended time periods, while the areas between were ravaged by ecological change. The relative stability of certain areas could be due, e.g., to topographic structures and edaphic conditions that affect the rainfall and soil humidity. Areas in the tropics able to maintain the rain forest through the periods of shifting climates, may also be optimal for maintaining unchanged ecological regimes in future situations with increased aridity caused by widespread man-made disturbance of the vegetation cover.

If this interpretation is right, the safeguarding of centres of endemism in the tropics would have implications far beyond the level of "cooling extinction hot-spots". To me, a montane forest with many endemic species should also be a top priority area for preventing disturbance of water catchment or other key elements of life-support for man. A montane forest with no endemic species (see Figure 1) would seem of lower priority in this respect: the lack of endemics signal that the ecological conditions of this area are naturally unstable.

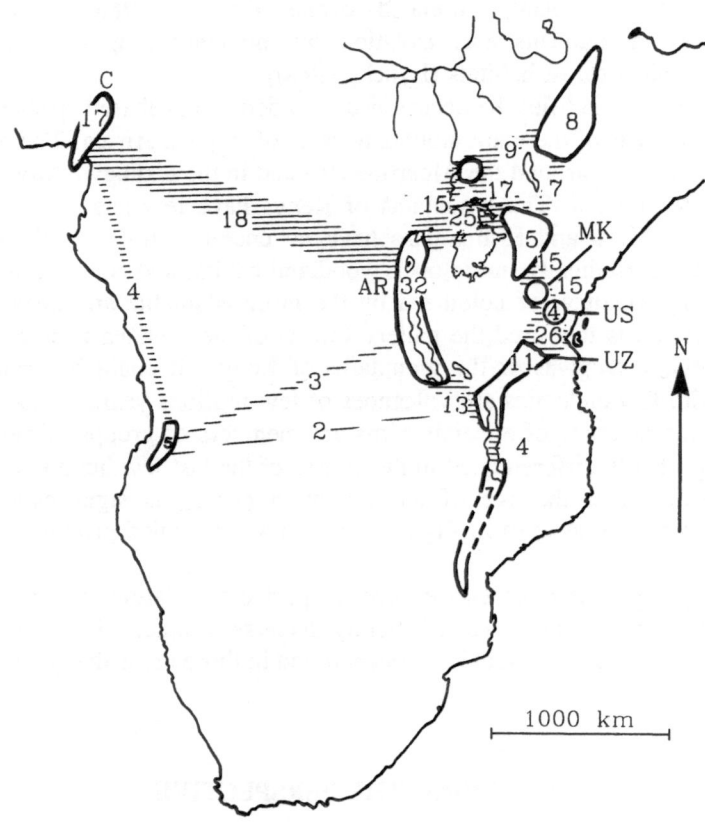

Figure 1. Relationships of the avifaunas of ten main mountain tracts in tropical Africa. Figures within the areas show the number of endemic species and semi-species of genuine high-elevation forest birds; figures between adjacent areas give number of shared taxa. AR = Albertine Rift, C = Cameroon Mts., MK = Mts. Meru and Kilimanjaro, Us = Usambara Mts., Uz = Uzungwa Mts. (Fjeldså, 1990; data mainly from Hall & Moreau, 1970).

I find it crucial to the planning of the protection of man's life support systems to test the durability of this prediction. Besides the basic methods of mapping distributions and biogeographic patterns, biogeographic research should also go into advanced speciation analysis. Key questions would be whether centres of endemism are just reflections of the present-day carrying capacity, or represent dispersal centres and, also, to what extent dispersal centres visible today represent areas that generated species periodically over longer geological time spans.

These questions are addressed in a long-term research programme planned in the East African montane forests by the Zoological Museum, University of Copenhagen. Testing might require comparison of palaeoclimatological data from assumed species-generating areas and those that lack endemic species.

THE MAPPING OF ENDEMISM IN PRACTICE

To date, no list of sites of global significance exists, despite this being perhaps the tool most urgently needed for planning conservation and development.

No measure of diversity can exist without a high degree of taxonomic evaluation and conformity. No plotting of centres of endemism can occur without a high degree of knowledge of the distributions of species. However, it would be madness to postpone the planning of conservation actions until the biodiversity is known in full. Shortcuts must be made, using data which already exist or which can be obtained by rapid surveys.

Plants represent the economically most significant component of the biodiversity, and therefore deserve strong attention (Prescott-Allen, 1988). However, only certain groups of vascular plants are known sufficiently well for rapid field surveys. We must therefore consider, if other groups may be more practical for initial surveys of patterns of endemism.

The consensus of a recent symposium over the Amazon rain forest, made by Conservation International and the Brazilian National Amazon Research Institute, was that a wide survey of key areas for preservation existed only for birds, with fairly good data also for frogs, monkeys, *Heliconius* butterflies and forest trees (Collar, 1990). Ehrlich (1988) suggests that the most useful groups for setting priorities are birds and butterflies: both are well-charted for detailed analysis and they have quite different roles, which reduces the preconception that could enter into judgments based on a single group. Bird data are best for global review, regarding detail and uniformity. For this reason alone, a compilation of existing data on endemic birds may become a cornerstone contribution for the 1990s. This is already in progress, as a writing-desk study (1988-92) made at the International Council for Bird Preservation and endorsed by UNEP (ICBP 1989, 1990).

WHY USE BIRDS?

The value of birds to set priorities for conservation is not at all universally accepted or understood. People involved in development and studying under-development will regard birdwatchers as fundamentalists defending their own special interest. Birds are marginal in relation to human self-interest in biodiversity. Nevertheless, ornithologists have obtained a role as front runners in the combat for biodiversity. This has several reasons:

1. Because of the Birds Themselves
Birds communicate by visual signals and calls, easily detected by man, and so are easy to study (unlike most mammals, which are skulkers communicating by smells). Belonging high up in the food webs, birds are sensitive to ecological change and, therefore, useful as ecological indicators.

2. Because of Ornithology
Ornithologists have found and described the earth's species, mapped their distributions and converted this knowledge into practical guides for further field-work more completely than students of other larger groups of organisms. Ornithologists have studied birds from a great variety of aspects: physiology, ecology, ethology and life history, and have been central investigators in the development of ecological theory. In museums, large collections of eggshells and skins provide comparative data for studying the historical change in environmental pollution.

3. Because of Popularity
In many cultures, birds represent aesthetical values and are significant as symbols of wilderness, and easily provoke defensive responses. To illustrate the power of this it suffices to remind of how Rachel Carson catalyzed the development of environmental awareness with a metaphor that referred to the absence of birds - Silent Spring (Carson, 1963). Because of this appeal, there is a large number of amateurs whose interest can be channelled into effective monitoring of population numbers and distributions, gridded distribution atlases (e.g., Blakers et al., 1984), continent-wide inventories of important bird areas (Scott and Carbonell, 1986, Grimmett and Jones, 1989) and index data showing population trends (Marchant et al., 1990). The world "amateur", here, is used in its more traditional sense: the person who does something for the love of it, rather than in the sense of doing it unskilfully. Because the "amateurs" care about the birds, they deliver enormous amounts of carefully collected data - at low cost. State agencies profit greatly on this.

For these reasons, birds play a central role as indicators of environmental change, e.g. as early warnings about the effect of insecticides (Carson, 1963, Ratcliffe, 1970) and other pollutants, the effect of changes in rural development, etc. This is well discussed in several textbooks, especially in Diamond and Filion (1987). Birds represent the key group of organisms used for selecting areas for nature protection in the EEC, and for wetlands (RAMSAR sites) globally.

The most recent development is ICBP's project to chart centres of endemism from bird data. Forerunners of this project were biogeographic analyses especially by Haffer (1974) and Diamond and Hamilton (1980), but the current ICBP project has more data available, profiting from a global network. The project (progress reports) compiles all known records of endemic birds, allocates density of endemism to gridded maps, identifies minimum sets of grid units needed to include all endemics of a region, and, by incorporating data about habitats and land use, shows how patterns of endemism may be affected by habitat loss. Other results include identification of general biogeographic patterns and comparisons of patterns of species of different habitat.

An Example: Avian Biodiversity in the High Andes

To exemplify the use of bird records for selecting areas for conservation I will use data from studies in the Andes (Fjeldså, 1987, Fjeldså & Krabbe, 1990: 846-7, Frimer & Nielsen, 1989).

The tropical Andean region and its foothills have the greatest concentration of biodiversity on earth. My study focused on the much neglected woodlands occurring above 3500 m elevation in the Andes. The highlands of Perú and Bolivia, known as puna, are dominated by steppe-like habitat, but with widely scattered woodlands. These relicts of a once far more widespread habitat dwindle away because of logging, grazing and burning, and consequent aridification.

Figure 2a shows the disjunct present occurrence of high-elevation woods in Perú and Bolivia. The field-work covered 260 sites above 3500 m and, in this easily surveyable habitat, 1 to 2 days of study seem to give fairly complete surveys of the resident birds in a study plot. Figure 2b shows higher diversity of birds in woodlands than in open land and considerable regional variation. In order to show more clearly the variation in conservation value, I gave each taxon a score (0-4) reflecting its geographic range and degree of differentiation (= unique adaptations to local conditions; see legend). Figure 2c shows summed "uniqueness scores" for all localities. The accentuation of regional differences, compared with Figure 2b, reflects a strong clustering of the more special birds in certain parts of the species-rich areas. A special analysis was used to reconstruct the evolution of the genuine high-elevation woodland birds (Fjeldså, in press). This indicates that three areas were particularly important to temporary entrapment of populations of once-widespread forms: the Cochabamba basin in the Bolivian Andes, the Apurímac Canyon and Cordillera Blanca, as well as the adjacent Pacific slope of Lima (names on Figure 2a).

The value of this exercise for conservation planning was tested by assembling species lists for the 10x10 km study area most rich in species, within the three suggested centres of evolution of species, each of these being studied for 3 to 6 days. The combined record for these three areas comprised 66 per cent of all landbird species established above 3500 m in Perú and Bolivia, 57 per cent of the "endemic" species, and 65 per cent of the species listed as endangered (Collar et al., manus). A random sample of 25 other combinations of three well-studied 10x10 km blocks containing high-elevation woodland in Perú and Bolivia gave, on average, 20.3 per cent of all species, 12.5 per cent of the endemics and 3.0 per cent of the endangered species. The contrast between these sets of figures illustrate (beyond my expectation!) how important it is to base the selection of sites for biological conservation on careful study and analysis. (To "hit" populations of all highland birds of Peru and Bolivia, a minimum of 25 localities are needed, according to current data).

I do not claim that 10x10 km areas suffice to maintain the biodiversity. Although some high-elevation forest birds seem to be able to survive for some time as exceedingly small populations, considerable buffer zones will be needed to secure the full biodiversity of the key sites. Starting with Diamond (1975), an extensive literature and several research programmes exist now applying the theory of island biogeography to nature reserves, seeing them as "islands in a sea of development". This has led to a consensus about the necessity of securing large areas (however, it is still controversial whether one large reserve is better than an aggregate of several small ones; see Simberloff & Abele (1976)).

ADEQUACY AND SHORTCOMINGS OF BIRD DATA

In order to know the usefulness of the ornithological approach, one should also examine how well patterns of avian endemism reflect more general biogeographic patterns.

Birds are often viewed as "feathered chariots" able to reach whatever patch of suitable habitat. Certain birds have been characterized as "supertramps". The notion holds true especially where seasonal or unpredictable climates have selected for dispersive or migratory behaviour. However, in fairly stable environments, where all suitable habitat will be almost constantly "saturated" with species, there is minimal premium on trying to explore unknown habitat. Many birds of tropical forest habitats have evolved a psychological blockage - a "fear of flying" - against crossing open land (Diamond, 1981). Although capable of flight, they are strongly sedentary.

In order to be certain about the value of the bird data, it is necessary to overlay data from other groups. So far, studies by Prance (1973, 1982), Simpson & Haffer (1978) and Mittermeier (1988), and a current consensus judgement of the ICBP

Figure 2 (opposite page). Bird diversity of high-elevation habitats in Peru and Bolivia. 2a shows the distribution of relict woodlands, especially of *Polylepis* trees at 3500-4800 m elevation, based on field surveys 1987-89 and analyses of maps and satellite photographs (shading = humid forest on lower elevation). 2b shows the total number of species in well-studied sites above 3500 m (data in Fjeldså, 1987 and unpublished for 1989, and Frimer & Nielsen, 1989). Open symbols: open habitat, filled symbols: woodland. 2c shows "uniqueness indices" calculated on basis of the data used for 2b. These indices are the sum of scores given to each species: no point is given to widespread and poorly differentiated species; 1 point to widespread species with moderate subspecies variation or monotypic species of somewhat restricted distribution; 2 points to species with a more marked geographic differentiation or a smaller range; 3 points to species with a small, disjunct or very patchy distribution, often with clearly differentiated local populations; 4 points to four very rare, endemic species.
(Fjeldså, 1990).

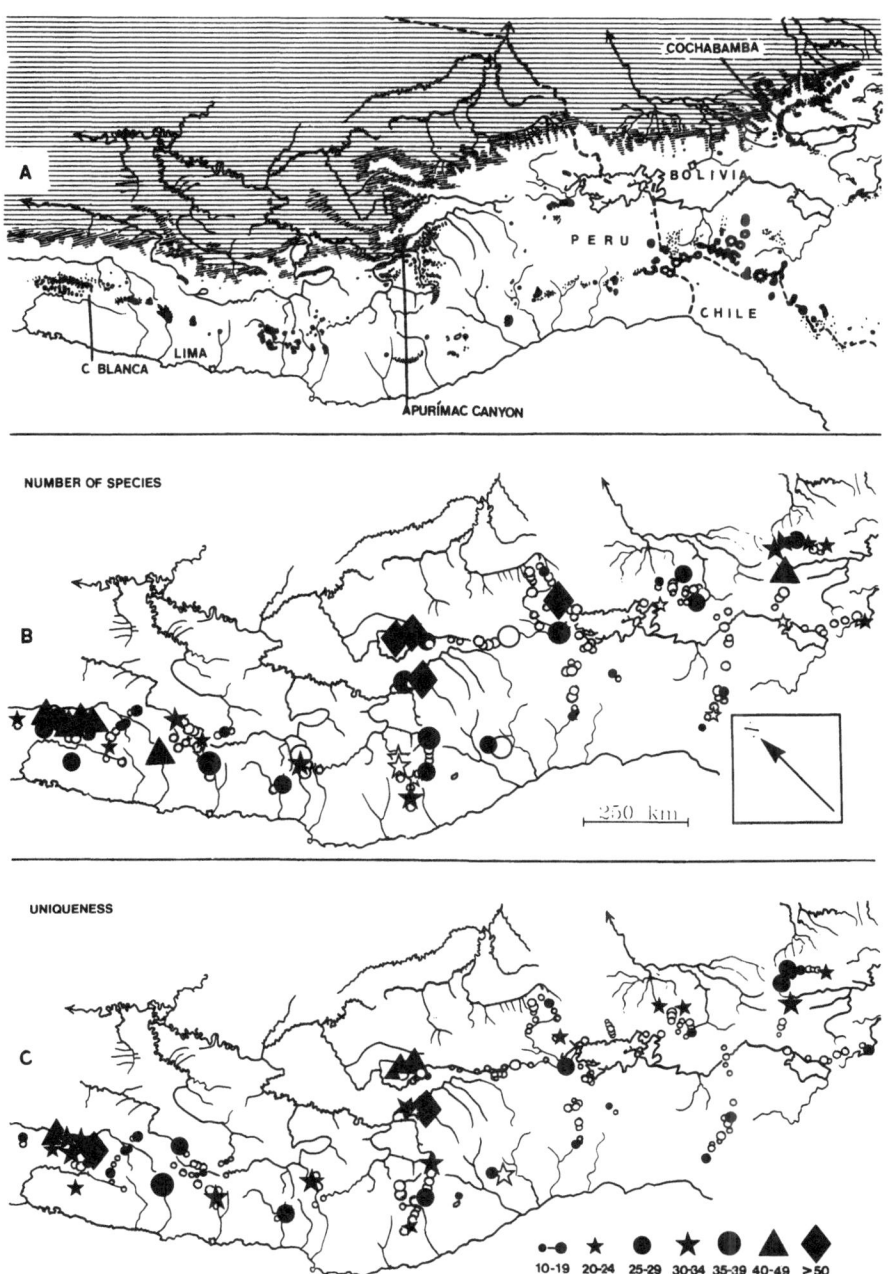

Figure 2. (See text on the opposite page).

biodiversity programme, indicate that whenever avian endemism is pronounced in the tropics, there is also a high degree of endemism in other life forms. Probably, the underlying speciation processes were similar, as well.

As seen in Figure 3, hypothesized Pleistocene forest refuges (Prance, 1982) and centres for endemic *Heliconius* butterflies (Brown, 1981) correspond well with the area for endemic lowland birds, but endemic upland birds extend slightly further inland (preliminary analysis, ICBP Biodiversity Project). In this area in southeastern Brazil less than 2 per cent of the original forest area remains (Scott and Brooke, 1985).

For high Andean forest and woodland habitat I compared bird distributions with published distributions of various groups of small mammals, insects and plants, and failed to see any general difference in extensions of distribution ranges or in patterns (however, statistical testing is difficult, because the birds have been mapped in greater detail).

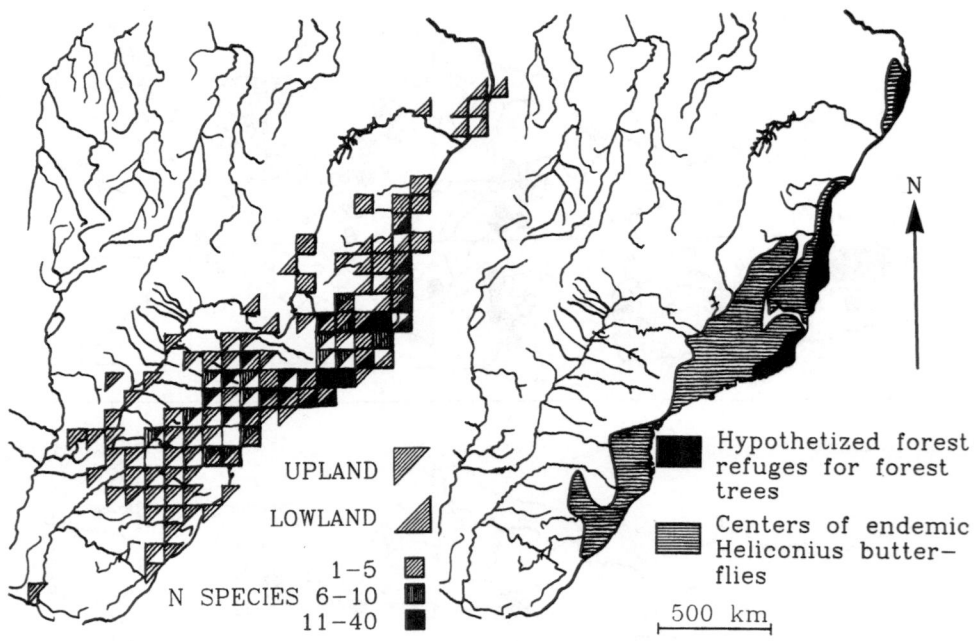

Figure 3. Southeast Brazil; a comparison of patterns of distribution: Endemic lowland and upland birds (left), based on mapping by T. Jones, ICBP headquarters, compared with centres of endemic *Heliconius* butterflies (Brown, 1981) and hypothesized Pleistocene forest refuges for tree families (Prance, 1982) (right). (Fjeldså, 1990).

While all avian centres of endemism may also have many endemics of other groups, birds will certainly not show <u>all</u> centres of endemism. Haffer (1969) suggested that larger organisms, like birds, evolve more slowly than smaller ones and, therefore, show less speciation. I suggest, as a possibly more important factor, that smaller organisms are able to survive in smaller patches of a quasi-stable habitat. It is well known that many plants maintain viable populations (surviving as endemic species) in geological formations (e.g., serpentine) of only a few hectares extension. Also, for many arthropods and land-snails, tiny pockets of special habitats may hold endemic species. Bird data do not reveal such tiny (past or present) refuge areas. Neither will they, in the same way as smaller organisms, reflect the biological uniquenesses of ancient lakes (Ochrid, Bajkal and Titicaca), or reflect the age of land features, such as erosion remnants of ancient mountain formations in the tropics. In plants, new (allopolyploidic) species may furthermore arise by hybridization, with no need for geographic isolation.

Apparently, bird data reflect the larger-scale habitat changes during the major climatic cycles of the Pleistocene properly, but fail on other points not yet fully defined.

IMPLEMENTING PRESERVATION

Cures, like causes, are proximate and ultimate. I propose securing biodiversity in three steps:

I. **Current biodiversity project**: Pointing out priorities using the present bird data and, also, studying congruence with available data from other groups of organisms. A strong focus on birds during this phase is a consequence of urgency.

II. **Using these data to plan further studies.** The first phase would reveal the most serious gaps in our knowledge of where to find areas with minimal conflict with development interests. This phase also includes speciation analysis and reconstruction of gross patterns of historical connection between montane forest areas. Such consensus may give the basis of priorities regarding where our knowledge should be expanded: where to secure a better geographic coverage and, also, where to supplement bird data with studies of plants, butterflies, etc., for finding finer patterns of endemism.

III. **The ultimate phase of implementing protection.** Irrespective of the motive for protecting species it is most meaningful to achieve this goal by securing the continued functioning of whole ecosystems. This widens the perspective in relation to human interest and, if successful, the survival of most of the species would follow, free of charge. Protected areas have been looked at as comforting delusions, which at best buy time (Guppy, 1984). Reserves must have considerable extension and they will never serve their intended function if the people surrounding them are landless and starving. The core of conservation areas must be surrounded by buffer zones used in sustainable manners, integrating conservation with economic and social development (Harris, 1984, Johns, 1985). The climatic consequences of deforestation (Jackson, 1989) mean that the question of nature conservation is inextricably bound to that of regional development in its fullest sense.

One of the key problems in reinforcing conservation is how to bring this message to the attention of the local inhabitants, or to the planners of "development". Prohibition (top > down) policy often fails and, in many cases, better results come by the education of local people (bottom > up). This encourages curiosity, interest and awareness of the importance of natural habitat. In certain cultures, the birds have proven to be valuable once again, when it comes to conservation awareness campaigns.

It is of utmost importance that funds are made available for monitoring that insures that the right choices have been made when we start dividing the world into refuges for biodiversity and production areas for man. For phase II in this plan, funds represent more critically limiting factors than man-power.

A comprehensive project (I-III) is a vast undertaking. Biologists must consider their funds for studies of biodiversity as start-up money. The long-term investment needed to solve the problems will, in most cases, be enormous. The time approaches when the conservationists should make integrated projects together with anthropologists, agricultural experts, etc., or should pass their pet projects on to the people in development agencies they once regarded as enemies (Stuart, 1986).

ACKNOWLEDGEMENTS

I wish to thank Peter Arctander, Niels Møller Andersen, Else Bering and Hanne Bloch for stimulating discussions during the preparation of the manuscript. I wish also to thank Nigel Collar and Timothy Johnson for updates regarding ICBP's biodiversity programme and for permissions to use unpublished maps for preparing Figure 2.

REFERENCES

Ab'Saber, A.N., (1982), The paleoclimate and paleoecology of Brazilian Amazonia. In Biological differentiation in the tropics (Prance, G.T., ed.), pp. 41-59. New York: Columbia Univ. Press.

Blakers, M., Davies, S.J.J.F., and Reilly, P.N., (1984), The atlas of Australian birds. Melbourne: RAOU and Melbourne Univ. Press.

Brown, K.S., Jr. (1981), The biology of Heliconius and related genera. Annual review of entomology 26: 427-56.

Carson, R., (1963), Silent spring. London: Hamish Hamilton.

Cody, M.L., and Diamond, J.M., (eds.) (1975), Ecology and evolution of communities. Cambridge, Mass.: Belknap Press.

Collar, N.J. (1990), Amazon as ark. World Birdwatch 12 (1-2): 10-12.

Collar, N.J., Gonzaga, L.P., and Krabbe, N., (manus), Threatened birds of the Americas. Cambridge, U.K.: ICBP and IUCN.

Collar, N.J., and Stuart, S.N., (1985), Threatened birds of Africa and related islands: the ICBP/IUCN Red Data Book, 3rd ed., part 1. Cambridge, U.K., ICBP and IUCN.

Connell, J.H., (1978), Diversity in tropical rain forest and coral reefs. Science 199: 1302-1310.

Diamond, A.W., and Filion, F.L., (1987), The value of birds. Cambridge, U.K.: ICBP (Techn. Publ. 6).

Diamond, A.W., and Hamilton, A.C., (1980), The distribution of forest passerine birds and Quaternary climatic change in tropical Africa. Journal of Zoology (London) 191: 379-402.

Diamond, J.M., (1975), The island dilemma: lessons of modern biogeographic studies for the design of natural reserves. Biological conservation, 7: 129-146.

Diamond, J.M., (1981), Flightlessness and fear of flying in island species. Nature 293: 507-508.

Ehrlich, P.R., (1988), Tomorrow's World: why saving biodiversity is today's priority. World Birdwatch 10(2): 6-7, 9.

Ehrlich, P. & A. (1983), Extinction. New York: Ballentine.

Erwin, T.L. (1983), Tropical forest canopies: the last biotic frontier. Bull. Ent. Soc. Am. 29: 14-19.

Fitter, R., (1986), Wildlife for man: how and why we should conserve species. London: Collins.

Fjeldså, J., (1987), Birds of relict forests in the high Andes of Peru and Bolivia. Copenhagen: Zoological Museum.

Fjeldså, J., and Krabbe, N., (1990), Birds of the High Andes. Copenhagen: Zoological Museum, and Svendborg: Apollo Books.

Fjeldså, J. in press. Biogeographic patterns for birds of relict woodlands in the high Andes. Acta XX Congressus Internationalis Ornitologicus.

Flenley, J.R. (1979), The rainforest: a geological history. London Boston: Butterworths.

Frimer, O., and Nielsen, S.M., (1989), The status of Polylepis forests and their avifauna in Cordillera Blanca, Peru. Copenhagen: Zoological Museum.

Grimmet, R.F.A., and Jones, T.A., (1989), Important bird areas in Europe. Cambridge, U.K.: ICBP (Techn. Publ. 9).

Guppy, N., (1984), Tropical deforestation: a global view. Foreign affairs 62: 928-965.

Haffer, J., (1969), Speciation in Amazonian forest birds. Science 165: 131-7.

Haffer, J., (1974), Avian speciation in tropical South America. Publications of Nuttall Ornithologists Club 14.

Haffer, J., (1985), Avian zoogeography of the neotropical lowlands. In Neotropical ornithology (Buckley, P.A., Foster, M.S., Morton, E.S., Ridgely, R.S., and Buckley, F.G., eds.), pp. 113-145. Washington D.C.: AOU (Orn. Monogr. 36).

Hall, B.P., and Moreau, R.E., (1970), An atlas of speciation in African passerine birds. London: British Museum (Natural History).

Harris, L.D., (1984), The fragmented forest. Chicago: Univ. Chicago Press.

Hamilton, A.C., (1982), Environmental history of East Africa. London: Acad. Press.

ICBP (1989), Biodiversity Project Progress Report No. 3. Cambridge, U.K.: ICBP.

ICBP (1990), Biodiversity Project Progress Report No. 4. Cambridge, U.K.: ICBP.

Imboden, C. (1987), The captive breeding controversy. World Birdwatch 9(3): 6-7.

Jackson, I., (1989), Climate, water and agriculture in the tropics. Essex and London: Longman.

Johns, A.D., (1985), Seléctive logging and wildlife conservation in tropical rainforest: problems and recommendations. Biological Conservation 31: 355-375.

Keast, A., (1985), Tropical rainforest avifaunas: an introductory conspectus. In Conservation of tropical forest birds (Diamond, A.W. and Lovejoy, T.E., eds.), pp.3-31. Cambridge, U.K.: ICBP (Techn. Publ. 4).

Lynch, J.D., (1988), Refugia. In Analytical biogeography (Myers, A.A., and Giller, P.S., eds.), pp. 311-342. London and New York: Chapman & Hall.

Major, J., (1988), Endemism: a botanical perspective. In Analytical biogeography (Myers, A.A., and Giller, P.S., eds.), pp. 117-146. London and New York: Chapman & Hall.

May, R.M. (1989), How many species are there on earth? Science 241: 1441-1449.

Marchant, J.H., Hudson, R., Potter, S.P.and Whittington, P. (1990), Population trends in British breeding birds. BTO, Tring.

McNeely, J.A., Miller, K.R., Reid, W.V., Mittermeier, R.A. & Werner, T.B. (1990), Conserving the World's biological diversity. Washington, D.C., World Bank.

Mittermeier, R.A. (1988), Primate diversity and the tropical forest: case studies from Brazil and Madagascar and the importance of the megadiversity countries. In Biodiversity (Wilson, E.O., ed.), pp. 145-154. Washington, D.C., Nat. Acad. Press.

Myers, N. (1983), A wealth of species. Storehouse for human welfare. Boulder: Westview Press.

Myers, N. (1984), The primary source. New York: W.W. Norton.

Prance, G.T., (1973), Phytogeographic support for the theory of Pleistocene forest refuges in the Amazon Basin, based on evidence from distribution patterns in *Caryocaraceae, Chrysobalanaceae, Dichapetalaceae and Lecythidaceae.* Acta Amazonica 3(3): 5-28.

Prance, G.T., (1978), The original evolution of the Amazon flora. Interciencia 3(4): 207-222.

Prance, G.T., (1982), A review of the phytogeographic evidence for Pleistocene climate changes in the neotropics. Annals of the Missouri Botanical Garden 69: 594-624.

Prance, G.T., and Lovejoy, T.E. (eds.) (1985), Amazonia. Oxford: Pergamon.

Prescott-Allen, R. and C. (1988), What's wildlife worth? Economic contributions of wild plants, and animals to developing countries. London: IIED.

Prigogine, A., (1988), Speciation pattern of birds in central African forest refugia and their relationship with other refugia. Acta XIX Congressus Internationalis Ornithologici: pp. 2537-2546.

Ratcliffe, D.A., (1970), The peregrine falcon. Calton: T. & D. Poyser.

Ricklefs, R.E. (1989), Speciation and diversity: the integration of local and regional processes. In Speciation and its consequences (D. Otte & J.A. Endler, eds.), pp: 599-622. Sunderland, Mass.: Sinauer Ass.

Rotenberry, J.T. and Wiens, J.A. (1980), Temporal variation in habitat structure and shrubsteppe bird dynamics. Oecologia 47: 1-9.

Salo, J., Kalliola, R., Häkkinen, I., Mäkinen, Y., Niemelä, P., Puhakka, M. and Coley, P.D. (1986), River dynamics and the diversity of Amazon lowland forest. Nature 322: 254-258.

Scott, D.A. and M. de L. Brooke (1985), The endangered avifauna of Southeastern Brazil: A report on the BOU/WWF Expeditions of 1980/81 and 1981/82. In Conservation of tropical forest birds (Diamond, A.W., and Lovejoy, T.E., eds.), pp: 115-140, Cambridge, U.K.: ICBP (Techn. Publ. 4).

Scott, D.A., and Carbonell, M., (1986), A directory of Neotropical wetlands. Cambridge, U.K.: IUCN and IWRB.

Scott, J.M., Csuti, B., Jacobi, J.D., and Estes, J.E., (1987), Species richness: a geographic approach to protecting future biological diversity. Bioscience 37: 782-788.

Simberloff, D., and Abele, L.G., (1976), Island biogeography and conservation: strategy and limitations. Science 193: 1032.

Simpson, B.B., and Haffer, J., (1978), Speciation patterns in the Amazonian forest biota. Annual Reviews of Ecology and Systematics 9: 497-518.

Soulé, M.E. (1987), Viable populations for conservation. Cambridge, U.K.: Cambridge Univ. Press.

Stenseth, N.C. (1980), Spatial heterogeneity and population stability: some evolutionary consequences. Oikos 35: 165-184.

Stuart, S.N. (1986), Usambara Mountains. World Birdwatch 8(3): 8-9.

U.S.Congress Office of Technology Assessment, (1987), Technologies to maintain biological diversity. OTA-F-330, U.S.Government Printing Office, Washington D.C.

Van Zinderen Bakker, E.M. (1982), African palaeoenvironments 18.000 yrs. B.P. Palaeoecology of Africa 15: 77-99.

Wilson, E.O. (1984), Biophilia. The human bond with other species. Cambridge, Mass.: Harward Univ. Press.

Wilson, E.O. (1985), The biological diversity crisis. Bioscience 35: 700-706.

World Commission on Environment and Development (1987), Our Common Future. Oxford and New York: Oxford University Press, 383 pp.

11.

MICROBIAL ECOLOGY AND POPULATION BIOLOGY

BRUCE R. LEVIN
Department of Zoology
University of Massachusetts,
Amherst 01003, USA

ABSTRACT

In their roles as decomposers, recyclers, pathogens, extractors and synthesizers, bacteria are responsible for maintaining the flow of nutrients through ecological communities, regulating the densities of populations in those communities and preventing the build-up of organic matter and toxic compounds. Microbial ecology and population biology study bacteria in these different roles and the factors affecting the distribution, abundance and evolution of populations of these microorganisms and their genetic parasites, plasmids, phage and transposons. I describe and critically consider three ways by which microbial ecology and population biology contribute to the quest for solutions to environmental problems; i) measuring and monitoring the effects of insults to the environment; ii) elucidating the factors responsible for maintaining the integrity of undisturbed ecological communities; and iii) identifying and developing microbes to degrade and recycle wastes, detoxify pollutants, as alternatives to noxious pesticides and as producers of useful compounds. I follow this broad excursion into the contribution of my field with a more personal consideration, confessions of an academic microbial population biologist. I conclude with a couple of caveats (musings); the liability of relying solely on technical solutions to environmental problems, and the effects of the separation of academic and applied biology on the quest for these solutions.

Key Words and Phrases: *Academic Science, Applied Science, Bacteria, Bacteriophage, Degradation of pollutants, Environmental monitoring, Genetic Engineering, Infectious Disease, Microbial Ecology, Plasmids, Population Biology, Population Ecology, Population Genetics, Risk Assessment, Toxic Wastes, Transposons.*

MY PROFESSION

My Profession Broadly Defined: Microbial Ecology and Population Biology

Bacteria and fungi dominate the final link in the food chain that commences with the fixation of solar energy; they are the decomposers responsible for preventing a continuous build-up of organic matter. Bacteria are also the great recyclers, they extract and make available the nitrates, sulphates and other inorganic compounds needed to maintain the nutrient flow cycle. Bacteria, along with other micro and not-so-micro organisms, are the parasites and pathogens that contribute to regulating the densities of, "higher", species. Bacteria are also the providers for those wanting, they synthesize or make available, vitamins, nitrogen and other compounds to organisms that are otherwise incapable of acquiring or synthesizing them.

Microbial ecology is the study of bacteria and other microorganisms in all of these guises (niches, if you prefer). It is an enterprise of extraordinary breadth. Whether they admit to it or not, a great deal of what microbiologists do; taxonomy, physiology, genetics, pathology, molecular biology, biochemistry, organic chemistry, and evolutionary biology, can be subsumed under the heading, "microbial ecology". Much of industrial and agricultural microbiology and a good deal of biotechnology can also be seen as microbial ecology. My field, microbial population biology, is one approach to studying this much grander picture.

Population biology includes studies of; i) the growth and regulation of population densities, ii) interactions between distinct populations (competition, predation, commensalism and symbiosis), and iii) the mechanisms responsible for maintaining genetic variability in populations and changes in their genetic structure, adaptation and evolution. It is an approach to the study of ecology and evolution in which populations of organisms (or somatic cells), rather than individuals, tissues or communities, are the object of study. Implicit in the population biological approach is the conviction that by elucidating the mechanisms that determine the genetic and ecological behaviour of single or few interacting populations, we will be able to understand the factors that regulate and determine the structure, stability and evolution of ecological communities and entire ecosystems. In essence, population biology is a reductionist, rather than holistic, approach to the study of ecology and evolution.

My Profession Narrowly Defined: What I do

While my formal training and most of my teaching is the genetics, population and evolutionary biology of higher organisms, for the past fifteen years my research has been almost exclusively with bacteria and the accessory genetic elements, virus (bacteriophage), plasmids, and transposons, that infect bacteria.

Bacteriophage (phage) are either, virulent (lytic), obligatory parasites (predators) whose replication is necessarily detrimental (usually lethal) to the infected host; or temperate, capable of reproducing by this lytic process but can also be transmitted vertically, in the course of cell division, by incorporating their DNA into the bacterial genome as prophage. Plasmids are autonomous molecules of DNA that are transmitted within cell lineages, vertically, or between cell lines, horizontally, by conjugation, or by being picked up by marauding bacteriophage, transduction. Transposons, "jumping genes", can only replicate when integrated into a bacterial chromosome, plasmid or phage (replicon) but can change locations within the bacterial genome by moving directly, conservative transposition, or by sending out copies of themselves, replicative transposition. Many plasmids and some transposons code for the machinery needed for their infectious transfer by cell-cell contact, conjugation. By virtue of their protein coats, bacteriophage can persist, without replicating, outside of their host cells. Plasmids and transposons are naked DNA, their survival as well as their replication requires their being within a bacterium.

Although their reproduction may engender a cost to their host bacterium, some temperate phage, plasmids and transposons carry genes that code for specific adaptations of their host cells. Included among these accessory element - determined characters are: i) resistance to antibiotics, heavy metals and ultraviolet light; ii) the capacity to degrade and utilize 'atypical' substrates (including some toxic pollutants); iii) the synthesis of allelopathic substances (bacteriocins and antibiotics); iv) the production of toxins, adhesions and other virulence determinants responsible for pathogenicity; and v) the synthesis of enzymes that degrade foreign DNA (restriction endonucleases). Thus, many of ecologically and clinically important characters expressed by bacteria are determined by genes carried on accessory elements, rather than chromosomes. Furthermore, as vectors for moving genes between bacteria of different (sometimes very different), as well as the same species, self-transmissible (conjugative) plasmids and bacteriophage also play a role in adaptation and evolution of bacteria. As a result of their capacity to move within genomes of bacteria and join segments of DNA, transposons generate genetic variability, the raw material of evolution. For general reviews of the basic biology and genetics of the accessory elements of bacteria see Falkow (1975), Broda (1979) and Campbell (1981).

Our research, (that of the students, postdoctoral fellows and collaborators working with me, as well as my own) includes both theoretical and empirical studies. The former involves the development and analysis of mathematical models of the population dynamics of bacteria and their accessory elements. The latter includes experimental studies with bacteria, mostly _E. coli_, and their plasmids, phage and transposons in laboratory culture, and genetic surveys of bacteria from natural populations. We have considered the conditions for the establishment and maintenance of the different kinds of accessory genetic elements in bacterial populations, their "existences" conditions, and the nature and consequences of (co)evolution in the host bacteria and these genetic "parasites", for reviews see Levin (1986, 1988), Levin and Lenski (1983,

1985), Condit, Stewart and Levin (1988). We have studied the genetic structure of natural populations of bacteria (Selander and Levin, 1980, Levin, 1981, Caugant et al., 1981) and the genetic relationship between pathogenic and non-pathogenic members of the same bacterial species (Porras et al., 1986, Plos et al., 1989). More recently we have been attempting to elucidate the ecological and genetic mechanisms responsible for the evolution of multiple antibiotic resistance plasmids (Condit and Levin, 1990), and the conditions under which evolution will lead to and maintain the virulence in pathogenic bacteria (Levin and Svanborg Edén, 1990).

MICROBIAL ECOLOGY AND POPULATION BIOLOGY AND THE QUEST FOR SOLUTIONS TO ENVIRONMENTAL PROBLEMS

Thus, be it narrowly or broadly defined, it would seem that the "environmental edge" encompasses my profession. This is certainly so, and practitioners of microbial ecology and population biology have been contributing to the quest for solutions to environmental problems in at least three ways; i) assessing and monitoring the consequences of insults to natural communities; ii) elucidating the mechanisms responsible for maintaining the ecological and genetic integrity of populations that comprise natural communities; iii) identifying and developing microorganisms to recycle wastes, detoxify pollutants, as pesticides and for producing alternative energy sources. In the following I describe and offer an occasionally critical view of the current status and utility of these different contributions of microbial ecology and population biology to the acquisition of solutions to environmental problems.

Assessment and Monitoring

What microbial ecology, population biology (and "Science" at large) does well is measure and develop instruments to measure. Given adequate time and facilities (really money and personnel) we can describe, with vast amounts of data and seemingly great precision, microbial communities and document the effects of environmental insults (e.g., chemical pollutants, temperature changes, pesticides, etc.) on those communities. We can ascertain, a posteriori, how these insults affect the distribution, abundance and genetic structure of the dominant populations (species) that make up these communities and the flow of nutrients and energy through them. With the application of the tools of molecular biology, this can be done for microbes that have not been formally classified by taxonomist as well as those that have never been cultured in the laboratory (Roszal and Colwell, 1987).

No matter how detailed, precise and impressive they may seem, by themselves, retrospective descriptions of the ecological and genetic consequences of environmental insults are of limited utility. They document the existence of problems and provide a rough idea of their magnitude, but they do not predict the extent to which the community will recover, the time recovery will take, or the consequences of its failing to return to its original state. By themselves, purely descriptive studies don't provide information about how to intervene to promote the recovery of the community, or restrict the magnitude of the insult. Retrospective studies are of little value for anticipating the environmental risks associated with the implementation of new technologies.

Elucidating the Mechanisms Responsible the Maintaining Integrity of Ecological Communities

For less-than-subtle environmental insults, like oil spills and the accumulation of toxic and non-degradable wastes, it doesn't require a great deal of ecological insight to predict the major, short-term, consequences of the insult and intervene to minimize those consequences. On the other hand, predicting the consequences of more subtle environmental affronts, like releasing novel organisms, and the long-term consequences of any insult requires an understanding of the mechanisms maintaining the distribution and abundance of the populations in undisturbed ("natural") communities, how those processes are affected by different kinds of perturbations, and the direction and consequences of evolution in those community. With this "understanding", we could design efficient (as opposed to exhaustive) programs to assess and monitor the health of communities and, ideally (but not necessarily), devise intervention schemes that will be effective in the long as well as the short-term.

Elucidating the ecological and genetic processes responsible for maintaining the integrity of undisturbed natural communities and the evolution of these communities is the traditional purview of academic (basic), as opposed to applied or mission-oriented, ecology and population biology. In the course of this century (and earlier) a great deal of effort has gone into the study of basic population and community ecology and population genetics. Most of this research has been (and continues to be) descriptive, providing the observations that have to be explained and, perhaps, offering hints as where those explanations may be lie. However, as can be seen from the perusal of a modern ecology and population biology text (e.g., Krebs, 1985; Begon, Harper and Townsend, 1986; Hartl and Clark, 1990), there have also been explanations; verbal, graphic, numerical, and mathematical models of the ecology and genetics of populations and the structure of ecological communities. These models identify the primary factors affecting the distribution, abundance, stability and evolution of populations that comprise natural communities. While their primary value may be heuristic, some of these models are useful for practical considerations, like designing programs for managing specific populations, e.g. fisheries (May, 1984). These "models" have also been employed to develop alternatives to traditional methods of agriculture, e.g., the control of plant pathogens (Barrett, 1979).

At this time, I believe our understanding of the mechanisms responsible for maintaining the integrity of natural communities of bacteria and higher organisms is too modest and too general to predict the long-term consequences of specific environmental insults and design programs to intervene to prevent or minimize those consequences. This limitation of ecological and evolutionary theory is illustrated by recent concerns about the consequences of releasing genetically engineered microorganisms. While we can readily generate scenarios about the potential ecological and genetic consequences of releasing specific types of "engineered microorganisms", there is little agreement among ecologists and other biologists about how realistic those scenarios are (see, e.g. Halversen et al., 1985). Worse than that, it is impossible to know whether the potential scenarios considered are exhaustive and difficult or impossible to actually test their validity (Simonsen and Levin, 1988).

I believe that in the long run our investment in the study of basic ecology and population biology will reap dividends in the development of solutions to environmental problems. (Of course he does, he's an academic biologist.) However, now and for the next few years to come, I expect the primary contribution of basic ecology and population biology to the resolution of environmental problems is going to be that of guidance; raising questions, pointing to where the answers may be found, and providing some of the tools to answer those questions.

Developing Microorganisms to Deal with Environmental Problems

Long before the existence of microbes was demonstrated and their ecological role determined, humans were already using microorganisms to clean-up the environment, recycle wastes and synthesize products; e.g., cesspools, compost heaps and fermentation processes. From one perspective, the build-up of solid wastes and toxic pollutants and our reliance on "dirty" fossil fuels and pesticides can be seen as a failing of our microbial clean-up, recycling and synthesizing mechanisms, or perhaps a failing of microbes themselves. If bacteria and fungi had the appetite, good taste and social responsibility, to eat and thoroughly digest all the delicious things we manufacture and dispose, to convert fixed solar energy into abundant clean fuels and to synthesize (or serve as) pesticides, the environmental problems we are confronting would be far less grave.

The identification, isolation and application of microorganisms and communities of microorganisms for disposing of wastes, recycling nutrients, decontaminating toxic pollutants, producing useful compounds and controlling pests has been a traditional endeavour of applied microbial ecology as well as industrial and food microbiology. These tasks are also fundamental to a good deal of contemporary biotechnology. In the more distant past, the microbes employed for these purposes were isolated from natural communities and recruited into human service with little intentional modification. More recently the effectiveness of domesticated microorganism has been improved by artificial selection. In both cases, the genetic variability being exploited already existed or arose by spontaneous or induced mutation. More recently, efforts to produce microorganisms for practical purposes have included the application of some of the more sophisticated tools of in vivo and in vitro (recombinant DNA) genetics to generate variation (Chakrabarty, 1979, 1986; Davis, 1988; Gartner and Kim, 1988; Malik, 1989).

The development of microorganisms to clean-up and recycle wastes, detoxify pollutants, produce or serve as alternatives to toxic pesticides and dirty fuels are, to my mind, the most important and most promising contributions of microbial ecology to solving environmental problems. These endeavours may also turn out to be the most important contribution of biotechnology at large. On the other side, the development of new organisms and technologies to employ them may, in themselves, be responsible for environmental problems that are even graver than those we are confronting currently. While I believe this is unlikely, it the unrestrained application of new technologies is the immediate cause of most of the environmental problems we are now confronting. Then again, the unrestrained employment of new technology will no longer occur in these environmentally enlightened times.

THE QUEST FOR SOLUTIONS TO ENVIRONMENTAL PROBLEMS: THE VERY MODEST CONTRIBUTION OF ONE MICROBIAL POPULATION BIOLOGIST

In the preceding section, I have considered how the fields of microbial ecology and population biology are contributing to the search for solutions to environmental problems. It seems appropriate to also address this issue from a more immediate and personal perspective, how these environmental problems affect my own professional activities; research, teaching and, for want of a better term, committee sitting.

While my profession may be perched on the environmental edge, environmental problems are, to a large extent, peripheral to my research. Most of the problems we study are those of academic ecology and evolutionary biology. The general goal of our research is to elucidate the processes that determine the distribution and abundance of populations of bacteria and their plasmids, phage and transposons, and the mechanisms of evolution in these organisms and genetic elements. Some of our investigations were motivated (and have been supported) by a specific environmental problem, concerns about the ecological and evolutionary consequences of releasing genetically engineered microorganisms. However, our approach to this more mission-oriented problem have been direct extensions of that we were already doing in our studies of the population dynamics of plasmids and gene transfer. These "risk assessment" studies provided us with an opportunity to evaluate the generality of our earlier research on these topics. We have not assessed the environmental and evolutionary risks of a specific program to release genetically engineered bacteria.

Thus, if our research is contributing to the quest for technical solutions to environmental problems, it is doing so only by providing information about, and hopefully increasing our understanding of, the processes that maintain the integrity of bacterial communities and the evolution of populations within those communities. Had the environmental problems not existed, I expect we would probably be doing the same research we are doing now.

As I indicated earlier, I teach basic genetics, population and evolutionary biology. Most of my courses are for advanced Biology undergraduate and graduate students, and focus on basic principles rather than the application of those principles. While in these "Major's" courses, we discuss some of the social and environmental implications of the basic science topics, these considerations are a small minority of the material treated. Save for occasional seminars, most of my pedagogical considerations of environmental and social issues are in the few lectures I give in a course entitled "The Biology of Social Issues", a large course for non-science students.

The Graduate students and Postdoctoral fellows working under my supervision do research on what seem, at least to me, as a very diverse array of topic. However, virtually all of these Graduate student and Postdoctoral projects are on aspects of the population biology and evolution of bacteria and their plasmids, phage and transposons. With the exception of the few studies associated with our genetic engineering risk assessment endeavours, their investigations, like my own, are motivated by problems in academic ecology, population genetics, evolutionary biology and epidemiology. We occasionally discuss environmental issues and how they are related to our research. However, these discussions are informal and usually occur while people are pipeting away or over lunch. Upon completing their work, the majority of the students and Postdoctoral fellows from my laboratory obtain jobs where they continue to teach and do research in academic rather than applied microbial ecology and population biology.

Occasionally, I put on my "Washington pants", adorn my chest with a brightly coloured strip of cloth, cover my torso in tweed, and board the 'Grey Pinstripe Express', the 7:30AM flight from Hartford to the "District", Washington D.C.. I am leaving the Ivory Tower for a brief excursion into the "real world". I have been asked to "serve", a ritual (perhaps right, but certainly a duty) of middle-aged scientists. For a few days, I will no longer be a professor plagued by recalcitrant students, contaminated cultures, bug-laden computer programs and postponed correspondence. I will be a "Consultant", an "Expert" (I used to wonder who they were), for some committee or meeting, in our nation's capital.

As I expect is the case in most countries, agencies of the United States government call upon university as well as government and industrial scientists for advice on environmental issues. This is also so for local and State governments and more enlightened industries. It has never been clear just how influential our advice and reports are to the actual formulation of environmental policy. [Is the formation of a "Committee-to-Study" the end or a first step?] On the other hand, it is clear that along with teaching, these committees, reports and occasional testimonies to legislative bodies, are our primary vehicle for dispersing information about environmental problems (spreading the faith) to those who could actually do something about those problems, other than study them.

CAVEATS AND MUSINGS

Indefinite Progress: Technical Solutions to Technical Problems
Progress is indefinite. There are technical solutions to all technical problems and these will always be found before their need becomes essential. When we finally exhaust our supplies of fossil fuel, we will have found abundant, safe, alternate energy sources. By the time the accumulated waste and toxic pollutants overwhelm us (to some, not many, they already have), we will have developed the technology to deal with them.

While few actually say these things, I expect many of us believe them, and virtually all of us want to. On the other hand, I don't believe there is justification for this kind of unbridled optimism. The perception of unlimited and timely technical progress is not much more than 200 years old, a very brief interval in the history of our species. The absolute reliance on technical solutions to environmental problems and faith in their existence may also exasperate the situation.

I believe that microbe-based technology will go a long way towards alleviating some of the problems of accumulating waste and toxic pollutants and provide alternatives to some of dangerous pesticides. On the other hand, it is critical that we don't fool ourselves into believing this technology will be our saviour. In the more distant future I expect we will able to specifically design and produce enzymes that degrade wastes and pollutants that we can't treat now. However, I believe that for the present and immediate future, we will have to rely on the existing enzyme-encoding genes and minor modifications of those genes for the production of degradative enzymes. There also could be synthetic compounds that will remain refractory to biodegradation, even with the development of true designer enzymes.

To be sure, the use of biological control (organisms) and "natural" (other-than human-made) products to control the pests and pathogens is more appealing than the largely synthetic (human-made) products we currently rely on for these purpose. However, there is no reason to believe these "natural" approaches to pest and pathogen control will extract us from the evolutionary arms race (pest and pathogen resistance countered by the development or evolution of alternative compounds and control agents) and other problems that plague current approaches to the these problems (NAS, 1986).

I believe it is critical that the scientists, engineers, planners and others involved in studying environmental problems and developing and applying technical solutions to those problems, publicly express as well as maintain humility about these endeavours. To be sure, it may be good politics to maintain a image of understanding and control and to promise quick technical solutions. However, I believe this image and process is irresponsible. It is clear that for the majority of environmental problems, the most effective immediate solutions are social rather than technical. Ultimately, these social solutions come down restrain, in our use of energy, manufactured goods, and agricultural products, especially by "first world". The illusion that environmental problems are understood and under control and the promise of soon-to-come, easy, technical solutions are hardly going to be an incentive for that restraint.

The Separation of the Academic and Applied Sciences

While the development of long-term solutions to environmental problems will require contributions of both academic and applied scientists, the communication between these two faces of science is limited. In virtually all universities in the United States (if not those in the world at large) there is a very sharp division between the basic and applied sciences. Administratively, pedagogically, and intellectually, academic science is separate from the applied science. Departments of Biology are distinct from those of Agriculture, Natural Resources, Engineering and Medicine. The separation between the basic and applied sciences is maintained at most research institutes as well as government and private agencies supporting research. Most critically, this separation is maintained in the journals. One publishes in either academic or applied journals and one tends to read the journals one publishes in.

The administrative, physical, pedagogical, and intellectual separation between the basic and applied sciences has a number of negative ramifications. It has established frictions between these otherwise complementary enterprises; that great human frailty, group identification, is pervasive. This separation has also had a profound influence on the training and attitudes of scientists as well as the subject and motivation for their research.

It is impossible to formally determine just how regressive the separation between academic and applied ecology and population biology is to the quest for solutions to environmental problems. On the other hand, I can't help but believe that it has retarded progress towards the development of those solutions. I am sure there are academic ecologists and population biologists who could make substantial contributions to the quest for solutions to environmental problems, if they knew enough about them. At the same time, I am sure there are applied scientist with the knowledge of those problems who are unaware of potentially useful contributions buried in the basic ecology and population biology literature. I am also convinced that the group identification has also been regressive, peer approval plays a major role in the choice of topics for one's research.

ACKNOWLEDGEMENTS

I wish to thank Simon J. Levin, Lone Simonsen, Susanne Bro and an anonymous reviewer for useful comments and suggestions. This endeavour was supported by grants from the US National Institutes of Health GM33782 and the US Environmental Protection Agency COOP No. CR-814309.

REFERENCES

Barrett, J.A. (1979), A model of epidemic development in variety mixtures. In Plant Disease Epidemiology. P.R. Scott and A. Bainbridge (Eds.), Blackwell Scientific Publ., Oxford, pp. 129-137.

Begon, M., J.L. Harper, and C.R. Townsend (1986), Ecology: Individuals, Populations and Communities. Blackwell Scientific Publications, Oxford.

Broda, P. (1979), Plasmids. W.H. Freeman, San Francisco.

Campbell, A. (1981), Evolutionary significance of accessory DNA elements in bacteria. Ann. Rev. Microbiol., 35: 55-83.

Caugant, D.A., B.R. Levin, and R.K. Selander (1981), Genetic diversity and temporal variation in the *E. coli* populations of a human host. Genetics, 98: 467-490.

Caugant, D.A., B.R. Levin, G. Linden-Janson, T.S. Whittam, C. Svanborg Edén, and R.K. Selander (1983), Genetic diversity and the relationships among strains of *Escherichia coli* in the intestines and those causing urinary tract infections. Prog. in Allergy, 33: 203-207.

Chakrabarty, A.M. (1979), Microbial genetic engineering by natural plasmid transfer - Some representative benefits and biohazards. In Genetic Engineering A.M. Chakrabarty (Ed.), CRC Press, Boca Raton, FLA., pp. 185-193.

Chakrabarty, A.M. (1986), Genetic engineering and problems of environmental pollution. In Biotechnology Vol. 8: Microbial Degradations, W. Schonborn (Ed.), Weinheim, FRG, pp. 515-530.

Condit, R., F.M. Stewart, and B.R. Levin (1988), The population biology of transposons: a priori conditions for maintenance as parasitic DNA. Amer. Nat., 132: 129-147.

Condit, R. and B.R. Levin (1990), The evolution of antibiotic resistance plasmids: The role of segregation, transposition and homologous recombination. Amer. Nat., 135: 573-596.

Davies, J. (1988), Genetic engineering: process and product. Trends in Ecology and Evolution, 3: S7-S11.

Falkow, S. (1975), Infectious Multiple Drug Resistance. Pion., London.

Gartner, F. and L. Kim (1988), Current applied recombinant DNA projects. Trends in Ecology and Evolution, 3: S4-S7.

Halversen, H.O., D. Pramer and M. Rogul. Eds. (1986), Engineered Organisms in the Environment: Scientific Issues. Amer. Soc. Microb. Eash. D.C.

Hartl, D. and Clark, A.C. (1990), Population Genetics Sinauer, Sunderland.

Krebs, C.J. (1985), Ecology: The Experimental Analysis of Distribution and Abundance. Harper and Row, New York, 800 pp.

Levin, B.R. (1984), Changing views of the hazards of recombinant DNA manipulation and the regulation of these procedures. In: Proc. Vth International Symp. on the Transfer of Antibiotic Resistance. V. Krcmry Ed. Avicenum, Prague. Republished in the Recombinant DNA Research Bulletin of the NIH.

Levin, B.R. (1986), The maintenance of plasmids and transposons in natural populations of bacteria. In: Antibiotic Resistance Genes: Ecology, Transfer and Expression. R. Novick and S. Levy Eds. Cold Spring Harbor Laboratory, pp. 57-70.

Levin, B.R. (1988), Frequency dependent selection in bacteria. Phil. Trans. R. Soc. London B 319: 549-472.

Levin, B.R. and C. Svanborg Edén (1990), Selection and the evolution of virulence in bacteria: an ecumenical excursion and modest suggestion. Parasitology. (In press)

Levin, B.R., and R.E. Lenski (1983), Coevolution in bacteria and their viruses and plasmids. In: Coevolution, D.J. Futuyma, M. Slatkin (eds.). Sinauer Associates, Sunderland, Massachusetts. pp. 99-127.

Levin, B. R., and R.E. Lenski (1985), Bacteria and phage: A model system for the study of the ecology and co-evolution of hosts and parasites. In: Ecology and Genetics of Host-Parasite Interactions, D. Rollison and R. M. Anderson (eds.) Academic Press. pp. 227-242.

Malik, V.S. (1989), Biotechnology - the golden age. In: Advances in Applied Microbiology, Vol.34. S. L. Neidleman (ed.), pp. 263-306.

May, R.M. (1984), Exploitation of Marine Communities. Proceedings of a Dahlem Conference. Springer-Verlag, Berlin.

NRC (1986), Pesticide Resistance: Strategies and Tactics for Management. U.S. Natl. Acad. Sci. Press, Washington D.C.

Plos, K., S. Hull, R. Hull, B.R. Levin, I. Ørskov, F. Ørskov, and C. Svanborg Edén (1989), The distribution of the P-associated pili (Pap), Region among *Escherichia coli* from natural sources: Evidence for horizontal gene transfer. Infect. and Immun., 57: 1604-1611.

Porras, O. D.A. Caugant, B. Gray, T. Lagerard, B.R. Levin and C. Svanborg Edén (1986), Difference in Structure between type b and nontypable *Haemophilus influenzae* populations. Infect. and Immun., 53: 79-89.

Roszak, D.B. and R.R. Colwell (1987), Survival strategies of bacteria in the natural environment. Microb. Revs., 51: 365-379.

Selander, R.K., and B.R. Levin (1980), Genetic diversity and structure in populations of *Escherichia coli*. Science, 210: 545-547.

Simonsen, L. and B.R. Levin (1988), Assessing the risks of releasing genetically engineered organisms. Trends in Ecology and Evolution, 3: S27-S30.

12.

INSECTS, MAN AND ENVIRONMENT: WHO WILL SURVIVE?

JOOP C. VAN LENTEREN
Laboratory of Entomology
Wageningen Agricultural University
PO BOX 8031, 6700 EH Wageningen, The Netherlands

I Believed Then

There was nothing I ever wanted more
than that village-outskirts green evening again
- I on my belly in the May grass with you
dress slightly drawn up over your thighs,
and may-beetles flying over us unsuspectingly,
death-doomed guests of the universe;
and I believed then that the world
would take me back again,
earth, trees, illusions surviving the winter,
that the world, tired of its losses, would take me back.

Sándor Csoóri.

ABSTRACT

Insects form the most diverse group of organisms, both in number of species (80 per cent of the animal kingdom) as well as in variation of biological characteristics and functions. Man has been interested mainly in two categories of insects: harmful and beneficial species. These two categories comprise only a few thousand of the millions of insect species.

A vast group of insects is classified as neutral, which is a rather trivial term because these species are essential components of the world's ecosystems, and thereby, contribute to man's long-term well being.

Before the development of synthetic pesticides around 1945, the role of entomologists in agriculture and medicine had been principally viewed as positive, but their accomplishments since then have been much debated. Although food production has increased and tropical vector-borne diseases have decreased because of entomological research, the manner in which these results were achieved - the production and application of environmentally hazardous biocides with the goal of exterminating pest species - was questionable, to put it mildly. Ecologists have voiced their concern almost immediately after the initiation of pesticide usage and continue to do so. Policy makers and politicians are now comprehending - almost 40 years after the first warnings - the environmental risks of pesticides and, accordingly, governments are currently changing crop and disease protection programmes.

The role of entomologists in helping to improve the environment is to be effected via the development of ecologically safer methods of pest and vector control by an increased use of nature's inventories (e.g. breeding of resistant plants and biological control). A more general issue of much greater priority is how changes can be effected on mankind's perspectives towards nature. Should we continue to exploit nature as we have done over the past three centuries, the extermination of mankind rather than of insects would be more probable, due to the inherently greater survival capacity of the latter.

Key Words: Chemical Pest Control, Entomology, Integrated Farming, Nature Attitudes, Non-chemical Pest Control, Pest Control Policy.

INTRODUCTION

To many people, the word *insect* is synonymous with the term, noxious organism. For example, insects are regularly mentioned as being harmful, cruel and dangerous creatures in books, magazines, movie and television shows. Upon asking someone to list some insect names, they will start giving such examples as ants, flies, wasps, locusts, and mosquitos. Beneficial, beautiful, or harmless insects do not usually occur among the first five species mentioned. Insects leading to pest situations are well known! Similarly, when asked about the role and importance of insects the same picture appears: long lists with negative characteristics. Only a few cite more positive aspects like pollination and production of honey and silk.

Therefore, I plea for studies resulting in a deeper understanding of insect ecology to be able to deal in a more balanced way with the problems which insects may create.

Most insects play a very positive role in nature while only a minimal fraction of insect species - less than 0.01 per cent - possibly cause problems for humans. Due to its activities in controlling this small group of pest insects - a development which started only 40 years ago - the entomological profession was blamed. Firstly, in this paper I will discuss the role played by insects and the problematic situation with regard to insect control. Then I will make suggestions on how to improve this situation. Finally, I will turn to a more general question and not limit myself to the field of entomology: I will contemplate on how mankind views nature and how this influences our future. I have structured the paper by formulating a number of statements.

Entomologists study insects, by and far the largest group of the animal kingdom. The number of animal species is estimated to be between 20 and 50 million out of which, to date, only 1.7 million have been identified and named. 80 per cent of all animal species are insects. Among the insects, plant eaters are our main competitors and they comprise about half of the insect world, i.e. some millions of species. The other half comprise insects which live off other insects (predators and parasites), insects which live off other animals and insects feeding on dead organic material.

Although the number of plant-eating insects is very large, the number of species regarded as pests is small. Some 2000 years ago, at the time of Pliny the Older (after Holland, 1635 in Southwood, 1977), the number of insects regarded as useful was equal to that noted as harmful. The 20th century books on applied entomology devote most space to the harmful species and often ignore, or overlook, the beneficial role of insects. Estimates of the number of pest species vary between 500 and 10,000 species worldwide. The last figure includes harmful insects which attack man or cattle, which damage man's dwellings or are generally a nuisance.

The category of medical and veterinary pests is not treated in my paper, although, particularly in this group, resistance to pesticides has developed extensively. Hence, alternative methods of control are of great priority to permit a stabilization or reduction of the substantially increasing number of human deaths resulting from vector-borne diseases, like e.g. malaria (*Plasmodium* spp.) transmitted by mosquitos (*Anopheles* spp.).

Even the most conservative estimate of the number of harmful insect species - the above mentioned 10,000 - makes it clear that very few insects species are problematic. Why is this number so small? This question brings me to my first statement:

MOST INSECTS ARE USEFUL, SO CONSERVE THEM TO PREVENT MORE PESTS FROM DEVELOPING

Ecologists have ample evidence relevant to the answering of the question formulated above: why are there so few pest insects? First of all, not all plant eating insects occur on the plants we produce for human use. 90 per cent of our food consists of only 300 plant species. Many insects feed on the other 300,000 plant species and are thus not directly detrimental to us. Furthermore, we rarely see these non-commercial plants being completely defoliated by insects. In nature, only 1 per cent of the plant biomass is eaten by insects and other herbivores. This is in strong contrast to the potentially 20 - 100 per cent consumption of food crops by insects. What has gone wrong? Comparison of population dynamics of potential pest species in natural ecosystems and agro-ecosystems points towards several general explanations with regard to the difference in degree of consumption.

- Wild plants possess resistance against a variety of insects, which results in no or slow population development.

- In natural ecosystems, host plants specific to certain insect species are not easily discovered, leading to high insect mortality during dispersal. Only a few insects are able to locate a host plant and reproduce.

- In natural ecosystems, a myriad of natural enemy species maintain plant-eating insects at low population densities and, even in agro-ecosystems, a number of potential pests is held at non-damaging levels by natural enemies which occur naturally.

These combined factors generally result in low population densities of plant-eating insects in nature. Agriculture brought the creation of monocultures of genetically identical crop plants, maintained in an excellent condition and often grown in the same geographical area over many years. This has led to the occurrence of maximal survival and reproduction rates for pest insects, whereas many species of their natural enemies have problems with surviving in monocultures. Thus, man has created first-rate favourable environments for plant eating insects through the provision of more accessible and/or more nutritious food or lush growth, as well as by the decimation of their natural enemies (Huffaker et al., 1976).

Extensive studies have been made to find out how the agro-ecosystems influenced pest population dynamics and how these situations could be changed to profit in a better way from the pest control mechanisms which nature provides freely (for a review see van Lenteren, 1987). In the framework of this paper it is important to mention that many beneficial insects occur in nature in the form of natural enemies of pest insects.

Besides beneficial insects as predators and parasites, we know other large groups of useful insects in the role as essential elements in keeping ecosystems functioning. Such groups comprise pollinators and recyclers (insects eating dead organic material), and smaller groups of beneficial insects producing nectar, silk, lac and wax (for a review see Smith et al., 1973).

Entomology has also played an important role in culture and recreation (e.g. Southwood, 1977; Hogue, 1987). Further, insects are used as food, although this was more important in the past than today. Large orthopterans are still an important item of the diet, especially to children, in certain parts of Africa (Southwood, 1977, for reviews of entomophagy see Bodenheimer, 1951 and DeFoliart, 1989).

ENTOMOLOGISTS SHOULD REDIRECT THEIR STUDIES TO IDENTIFY LONG-TERM, ENVIRONMENTALLY SAFER SOLUTIONS OF PEST CONTROL

To combat the relatively few pest species, some 200 different chemical ingredients are used in an array of formulations (Besemer, 1985). Insecticides form the most hazardous category of the pesticides because, unlike fungicides, they are aimed at killing animal life. The majority of insecticides can be characterized as having a broad-spectrum activity, with well known risks for producers, appliers, consumers and the environment. These risks are of general concern.

The main problem for the chemical industry at present, is however, the development of insect resistance against pesticides. The exponential increase of resistance leads to a dramatic rise in human disease problems (e.g. malaria, due to insect-vector resistance) and a decrease in the yields of some crops (Metcalf, 1980; Dover and Croft, 1984). Furthermore, the development of new pesticides has become increasingly difficult. As many more potential chemicals need to be screened, the overall production costs are rocketing and more research is necessary before new pesticides are legislated. The rate at which insects are developing resistance to new and complex pesticides is, however, not decreasing. Chemical pest control has resulted in more than 500 insect species becoming resistant to one or more pesticides. All attempts to eradicate pest insects have failed. Harmful insects survived all tactics we have invented in order to destroy them.

The above factors, combined, will lead to ever increasing costs of chemical control. As a result, a dramatic decrease in the number of newly marketed insecticides appearing per year has already been experienced over the last two decades: 20 new active ingredients were registered yearly in the sixties, which is in strong contrast with the less than one being registered per year at present. In relation to the problems just mentioned, it is my opinion that the role of agricultural entomologists in pest control will have to change.

Since the Second World War many entomologists have been dealing merely with the technical problems of developing, testing and applying insecticides. Much of the information available on the biology of the pest organisms concerned have remained unused. Development of ideas on how pests originate and how this may be prevented did not seem necessary when cheap and powerful chemical pesticides were available.

Actions which are aimed at the control of individual species, will result in new problems if studies are not done in an holistic ecosystem approach. Inconspicuous, but essential changes in the functioning of ecosystems are often only perceived over many years. Most professional entomologists now derive their livelihood by developing chemical methods to kill insects, or through the training of younger colleagues who will do this job for them in later years (Odhiambo, 1977). I am convinced that the moment has arrived to change from an opportunistic method of (chemical) pest control - whereby problems are awaited, solutions are hoped for and the afore-mentioned indicators of resistance development and the low number of new pesticides available are ignored - to a biologically based method of pest control. Many alternative methods are already available (Table 1).

One of the powerful alternatives available is biological control and, if one compares statistics related to the development and application of chemical and biological control, one wonders why biological control is not applied on a substantially larger scale. Both the success ratio of new agents and the economic data show that biological control can be much more profitable than chemical control (van Lenteren, 1986). Further, the effects of natural enemies on the environment may be neglected, while such effects tend to be of major concern in chemical control. In my view, the slow increase in the use of biological control is more a matter of politics and policy than of any inherent weakness of the method itself. This also applies to several other methods mentioned in Table 1 which have been developed but receive too little support to be implemented.

APPLICATION OF NON-CHEMICAL PEST CONTROL IS NOT STIMULATED

During the past three decades many countries have invested public money in the development of non-chemical control methods. In this section I will try to identify the reasons why so few of these methods have been extensively used.

Farmers' Attitudes

Until very recently, only few farmers (organisations) asked for, or stimulated, development of non-chemical control methods. The adoption of insecticides was rapid because they allowed the farmer to decide when and where they should be used. Decision criteria were clear: the method was easily understood, it was effective (at least in the short term), reduced labour costs, and was a practice which the farmer could control and decide upon independently of his neighbours, institutions or agencies. Initially it was a straightforward technology.

In contrast, biological and integrated control are more complicated because of the requirement for monitoring of various pests, integration of different control methods and situation specific prescriptions. The latter systems require a degree of knowledge and sophistication much greater than the one demanded by pesticide technology (Perkins, 1982).

Table 1
Methods to prevent or reduce the development of pests.
(van Lenteren, 1990).

Prevention:

- prevent introduction of new pests (inspection and quarantine)
- start with clean seed and plant material (thermal disinfection)
- start with pest free soil (steam sterilization and solarization)
- prevent introduction from neighbouring crops.

Reduction:

- apply cultural control (crop rotation)
- use plants which are (partly) resistant to pests
- apply one of the following control methods:
    - mechanical control (mechanical destruction of pest organisms)
    - physical control (heating)
    - control with attractants, repellants and antifeedants
    - control with pheromones
    - control with hormones
    - genetic control
    - biological control
    - (selective) chemical control.

Control based on sampling and spray thresholds: guided or supervised control;
Control based on the integration of methods which cause the least disruption of ecosystems: integrated control.

Initiatives concerning the development of biological control programmes have - up to now - been taken by researchers and policy makers and they must continue to do so. Being unable to control a pest with chemicals is a stronger reason for farmers to change their ideas of biological control than ideological reasons. As soon as farmers realized that chemical control was no longer sufficient for complete control, their interest in an integrated approach was generated. We should not reproach the farmer, because governments legitimate the use of chemicals and often state that when chemicals are used as advised, they do not contaminate food or the environment and do not harm plants, animals or humans (van Lenteren, 1987).

Currently, the attitude of several groups of farmers is changing. Dutch fruit growers and producers of greenhouse vegetables, for example, have experienced the positive aspects of integrated control and seriously worry about the increasing public concern about pesticide usage. Therefore, at present they generally prefer to use biological control methods (van Lenteren and Woets, 1988). Still, the most effective way to reduce usage of insecticides is through modification of the pesticide legislation policy, through stimulative measures (subsidies for environmentally friendly techniques) or fines (for incorrect use of pesticides).

The Viewpoint of the Chemical Industries

In general, we can state that any complication in a simple, pure chemical pest control programme is appreciated as a negative development by the large industries. Biological and natural control, which are regarded as cornerstones of integrated control, not only complicate chemical control programmes, but they seem to be commercially unattractive as well because of a combination of (van Lenteren, 1986) the following:

a) the impossibility to patent natural enemies,
b) complicated mass production,
c) short shelf-life,
d) specificity (too small market), and
e) different and more complicated guidance for growers.

Classical biological control may never interest the industry, because beneficial insects are introduced only once and mass production is not necessary. Programmes requiring mass production and frequent releases, such as the programmes applied in large scale greenhouse facilities e.g. in The Netherlands and the UK, offer better possibilities. At present, the industry shows some interest in microbiological pesticides (e.g. insect-killing fungi, viruses and bacteria) as they are much easier to produce, store, ship and apply than parasitic and predatory insects. Presently, this causes problems, because the specific interest in these microbial pesticides by the large industries might generate the idea with farmers and policy makers that they are potentially better natural enemies than predators and parasites, and this is certainly not the case (van Lenteren, 1983).

Chemical industries will not start production of other than broad spectrum pesticides on their own initiative, unless the use of those pesticides is prohibited or when pest organisms substantially develop resistance - but time is on our side! We cannot blame the chemical industry for this attitude, because their goal is to make a profit. The industry provides pesticides which are allowed for use by a government's legislation and registration policy.

Role of the Governments
Therefore, it is the governmental bodies who should take the lead here. They are in fact the only ones able to change the pest control picture through measures that make some kinds of chemical control less attractive or impossible to use (by measures concerning registration, taxation, side-effect labelling etc.), and by stimulating other control methods (by funding research, but above all by teaching on all levels in order to change the attitude towards nature, and by improvement of the extension service). It is a rather bizarre situation that public money is used for the development of alternatives of chemical control when, at the same time, their application is often not encouraged due to the overall presence of (too) cheap, broad-spectrum pesticides.

CHEMICAL CONTROL IS IRRESPONSIBLY AND UNREALISTICALLY CHEAP

On the average the costs of pest control are very low and amount to only 3 per cent or less of the investment for producing a crop (van Lenteren, 1990). The costs of pesticides are in part so low because the chemical industries have - until recently - not been held responsible for any indirect costs inflicted upon the environment or man (e.g. for cleaning up chemical wastes or for reduction in diversity of flora and fauna). The availability of very cheap pesticides has resulted in a decrease of treatment thresholds. This makes the application of other methods than chemical control harder than before these chemicals were on the market and more damage had to be tolerated.

Appreciation of the indirect costs (social and environmental) and inclusion of these costs in their sales price results in more realistic costs of the use of pesticides and will automatically lead to an increased probability that alternatives will be used. However, the extreme complexity of estimating indirect costs related to pesticide use and the scarcity of data have delayed an assessment until recently. The American entomologist David Pimentel and his co-workers have been pioneers in this area (see e.g. Pimentel et al., 1980) and some of his information is given below.

It is important to realize that weeds, diseases and pests - despite the use of pesticides - annually destroy 33 per cent of the production in developed countries (USDA, 1965; Pimentel et al., 1978), and that this percentage is on the increase as a result of insect resistance to pesticides (see section 3). If pesticides were not used in the U.S.A, crop losses would rise by 9 per cent or 10,900 million US$. Preventing this loss requires an investment of 2,800 million US$ (Pimentel et al., 1978a). This gives a return of 4 US$ per 1 US$ invested, if only the direct costs of pesticide use are taken into account. Pimentel et al. (1980) estimated the annual indirect costs of pesticides to be 840 million US$ (Table 2). Adding this to the direct costs of 2,800 million US$ results in a return of 3 US$ per dollar invested in pesticides, rather than the 4 US$ given above. Pimentel et al. (1980) warn against the obvious over-simplification of this estimate. How can one place a monetary value on human lives lost or disabled due to pesticide poisoning? What is the price of wildlife losses? Some environmentalists state that the estimate of 840 million US$ is only a small portion of the environmental and social costs. A more complete cost/benefit analysis would therefore further reduce the profitability of pesticides and lead to a better appreciation of the usefulness of other control methods.

A similar estimate of indirect costs was recently made for pesticide use in The Netherlands (van der Vaart, 1987). Agriculture in The Netherlands is much more intensive than in the USA, which is reflected in direct costs of pesticide application: respectively 175 US$ per ha in The Netherlands and 9 US$ in the USA As a result of the higher pesticide load, a higher population density, more limited possibilities for recreation, smaller natural areas and a higher intensity of ground water used for drinking water, the indirect costs of pesticide usage will undoubtedly be higher in The Netherlands. The direct costs related to pesticide use in The Netherlands are estimated to be 275 million US$ per year (Nefyto, 1990). The estimate of indirect costs amounts to 275 million US$, but this is underestimated, because it does not include all the elements of Pimentel et al.'s (1980) estimate. Thus, for countries with a very intensive agriculture as in Western Europe, the indirect costs of pesticide use are much higher than for countries with a less inten-

Table 2
Total estimated annual environmental and social costs
of pesticides in the USA (After Pimentel et al., 1980).

Factor	Total costs (US$)
Human and animal pesticide poisonings, and contaminated livestock products	196,000,000
Reduced natural enemies and pesticide resistance	287,000,000
Honey bee poisonings and reduced pollination	135,000,000
Losses of crops, trees, fish and wildlife	81,000,000
Government pesticide pollution controls	140,000,000
Total	839,000,000

sive agriculture. As a result of this, the cost/benefit ratios are much lower than calculated on the basis of direct costs only.

A re-evaluation of the cost/benefit ratios for crop protection methods is essential to obtain a change in crop protection policy and a more realistic appraisal of the usefulness of alternatives.

THE DANGER OF A REDUCTIONIST APPROACH:
Overspecialization in Agriculture Prevents the Development of Agriculture with a Lower Pesticide Input

Agricultural research has seen a large diversification in this century. In plant breeding the search for better yielding varieties has been the central theme for years. Plant production concentrated on finding and overcoming limiting factors in plant growth and crop yields. Plant resistance against pests was not until very recently considered to be important, because most pests could easily be controlled with insecticides. This has led to plant varieties becoming less resistant to insects. In addition, over-fertilization resulted in plants which stimulate pest development (e.g. aphids, leafminers, spider mites).

Particularly during the past 50 years, pest control has been regarded as an independent area in agriculture. The viewpoint was that those working in this area could easily present solutions to protect the maximally productive crops. This reductionist approach to agricultural research has resulted in negative feedbacks leading to an increased use of pesticides. The situation is already changing: pest control is no longer seen as a completely independent subject, now that we know that solutions in chemical control are not unlimited and that pesticides cause serious side-effects. Still, much remains to be done and a more holistic approach, through a strong integration of the research areas of plant breeding/production and crop protection, will prevent developments in one area which may lead to the development of environmentally unattractive activities in the other.

BACK TO A HOLISTIC APPROACH:
Integration of Plant Production and Plant Protection Studies Results in Reduction of Environmental Problems

A first step towards the reduction of environmental problems is to test new developments within the framework of a farming systems approach. Entomologists, plant pathologists and weed specialists would work together with agronomists and plant breeders in such an approach. The focal point in the farming systems approach is farm economics in the form of **maximizing net income**, which is not synonymous with yield maximization. Top yields are obtained with excessively high inputs of fertilizers and pesticides. Reducing the inputs may lead to somewhat lower yields, but financial inputs are also lower and the net income may be the same or even better, as I will illustrate below. In farming systems, ecological and environmental effects such as pollution of soil and water by pesticides, can be minimized. In general, integrated farming takes more completely into account the various impacts on ecosystems (preservation of flora and fauna, quality and diversity of landscape, and the conservation of energy and nonrenewable resources) as well as sociological considerations (employment, public health and well-being of persons associated with agriculture) than is the case with current farming (Vereijken et al., 1986).

Although the objective of obtaining a better balance between the various influences - based on a growing awareness of problems caused by conventional farming - is often expressed, its realization is difficult to achieve as long as politicians and policy makers merely pay lip service to integrated farming and do not provide the means for its accomplishment.

In Europe, the farming systems approach is studied e.g. in the Federal Republic of Germany and The Netherlands. The aim is to investigate comparatively the technical and economical potential of low input systems (e.g. so-called integrated farming, or organic types of farming, with great environmental benefits) and high input systems (so-called present-day or current farming with low environmental benefit). An integrated farming system is: *"a coherent farm unit or set of units which relies basically on cultural and biological inputs, with chemicals as integrated supplements"* (Vereijken et al., 1986). The main aims are: *"to minimize inputs of nonrenewable resources and to provide a better balance between adequate production of yields and farm income on the one hand and ecological, environmental and sociological aims on the other. All these considerations must be compatible with cost effectiveness."* An important new element of such integrated farming studies is that they might provide a long-term alternative for current farming which leads to overproduction and environmental problems.

A description of such a programme can be found in Vereijken et al. (1986) and Zadoks (1989). Ideally three systems should be compared:

1. High input system: exclusively aimed at maximizing short-term yields and profits
2. Integrated system: aimed at optimum yields in terms of economics, long-term stability and environmental quality
3. Low input system: aimed at optimum use of natural elements and processes and where synthetic inputs are as low as possible (e.g. organic farming).

The practices which can be manipulated in such programmes are crop rotation, cultivation, fertilization, pesticide use, cultural control measures, biological control and other alternatives to conventional chemical control. Studies on demonstrational research farms in The Netherlands, although only recently begun, have shown the possibility of economically successful integrated farming and have led to many new cultural practices. Some of the practical results obtained in one of these projects at Nagele, The Netherlands, are summarized in Table 3.

Integrated farming gives slightly better economic results than present-day (= current) farming, whereas organic farming (in this case biodynamic farming, see Zadoks, 1989) is much less successful in this respect. The generally lower physical yields of the integrated system were compensated by a cost reduction as a result of the lower input of pesticides and fertilizers. Indirect costs of fertilizer and pesticide use are not yet included in this comparison. Integrated farming would give an even better result if indirect costs were included.

Table 3

Overall farm inputs (labour, fertilizer, pesticides) and percentage return of costs of three farming systems over the years 1985-1987; present-day = current farming is set at 100% (pers. comm., F.G. Wijnands).

	Current	Integrated	Biodynamic
Labour	100	100	200
Fertilizer	100	78	60
	mainly anorganic	75% organic	organic
Pesticides	100	9	0
% Return of Costs	72	77	67

I realize that I have strongly emphasized the economic aspect of integrated farming. I have done this, because economic figures are more convincing to farmers, policy makers and politicians on the short-term, than pure environmental reasons. In my own opinion development and application of sustainable agriculture, even if it were currently more costly, should have the first priority in Europe.

In integrated farming an important reduction of environmental pollution is realized through reduction in fertilizer use and the replacement of chemical pesticides by intensified knowledge of nonchemical measures (crop rotation, use of resistant varieties). In current farming systems mainly artificial fertilizers are used, whereas in integrated farming these tend to be replaced by organic manure. The total amount of N is lower in integrated farming to prevent excessive plant growth, resulting in a higher sensitivity to pests and diseases. Pest control in current farming is chemical. Weed, pest and disease problems are reduced in integrated farming through the use of weed competitive or disease and pest resistant varieties, reduction of N-fertilization, adaption of a specific sowing date and plant spacing, mechanical weed control, natural control etc. Chemical pest control in integrated farming is based on pest population sampling and use of decision thresholds. The very large reduction in pesticide usage in integrated farming - for a large part - is the result of using nematode resistant plants, thereby making soil disinfection against nematodes redundant.

The results on the demonstration farms have been so encouraging that the implementation of integrated farming is being strongly supported by the Dutch Ministry of Agriculture, in order to reduce environmental pollution consequent of agricultural activity and to create a firmer basis for the survival of agriculture in the longer term.

THE ENVIRONMENT SHOULD BE THE CENTRE INSTEAD OF THE EDGE OF THE ENTOMOLOGICAL PROFESSION

The above statement is a general one and applies to all professions where environmental problems play an important role. Entomology, through chemical pest control, contributes to environmental problems. To solve the environmental crisis, simple technological solutions are no longer sufficient (see e.g. the report from the World Commission on Environment and Development: "Our common future", 1987). The world-wide environmental problems demand a change in attitude towards nature, will any policy aimed at improving the environment lead to success.

Governments design and implement environmental policies, but no governmental policy succeeds without a broad social support. This means that a certain basic attitude towards nature is essential in the first place to develop ideas for an appropriate environmental policy and, secondly, to have the policy accepted and practised by the general public (Zweers, 1989).

Where have developments concerning basic attitudes towards nature gone wrong?

I am neither able, nor is it my primary task, to summarize the large diversity of basic attitudes and give an adequate historical review. On the other hand I cannot simply ignore this important topic. Therefore, I will reflect on the changes in attitudes towards nature which relate to my profession.

Philosophers have published on ethical, metaphysical, culture philosophical and social political aspects of the environmental problem. The problematic relationship between man and nature forms the focal point of these publications (e.g. Passmore, 1974; Rodman, 1983; Zweers, 1989). The basis for different attitudes is the perception or view one has about nature. The different attitudes also inform us about the individuals supporting them: how they see themselves, what is important for them, and what role they think they can play in nature. The full spectrum in these attitudes is expressed from acceptance of dominant western norms and values, to a radical rejection of these. Views are for a large part built on how each individual's judgement about the instrumental or intrinsic value of nature. A large variety of attitudes is distinguished by philosophers. I will only present three attitudes here to indicate the breadth of distance to be bridged between the groups adhering to these attitudes.

Man as Participant

Historically man has played a role as participant in nature. Like other organisms, he used and was used by nature in the form of nutrient and energy cycles. He participated within the ecosystem as one of the many components of the system. Man is merely part of the system which will also continue to exist when man is absent.

(The modern vision of the participant attitude contains several additions. The historic participation attitude is merely a biological participation. Besides biological participation, modern participation demands recognition of the intrinsic value of nature. The typical elements of humanity (e.g. culture, ethics, science and technology) are not focused on, or primarily used for submission of nature to man, but are used to create a situation in which one participates in nature.)

Man as Despot and Ruler

Since the early days of agriculture man has gradually left the role of participant. Local overexploitation and pollution interfered with natural cycles. As soon as man developed the idea that he was allowed to manipulate nature, nature was at risk. Cicero (106-43 B.C.) is one of the first to present a strict anthropocentrical view of nature ("De Natura Deorom"): *"everything in nature exists to the benefit of man, nature is created for man."* As far as my profession is concerned, it is interesting to note that the idea was expressed during the same period that it is man's task to eradicate those animals threatening the human world. (This same view was repeatedly used throughout this century to motivate extensive chemical control campaigns). It was not until the seventeenth century that anthropocentric views were widely accepted and the destruction of nature commenced.

Bacon (1561-1626) wrote "The New Atlantis" (1626) in which he designed a technocratic utopia partly realized during the following centuries. In "The New Atlantis" science and technology have no limits, nature has been eliminated and made redundant by a man-made, strongly improved environment: larger and sweeter fruits are produced than nature knows, the climate is largely under control, new plants and animals are created and life is artificially prolonged. Several European scientific societies emerging in the seventeenth century, with the English Royal Society as the first, had as their principle goal to rule nature through science and technology, and referred to Bacon's New Atlantis. The unlimited tendency to control nature, which Bacon described as strongly desirable, has led to our environmental crisis: his utopia has changed into a nightmare (Achterhuis, 1990).

Descartes (1596-1650) was reasoning along the same lines as Bacon. He sees man as the lord and master of nature, which, through technical inventions, should be submitted to him. A dead, quantitatively measurable, machine-like nature without intrinsic quality, envisioned by many seventeenth century scientists and philosophers, replaced the earlier visions of a living organic nature ("mother earth"). Until the seventeenth century the western attitude towards nature was similar to those in other cultures: mother earth did not demand managing and submission, but carefulness and respect with the role of man as participant only.

Since Bacon's and Descartes' time, nature is often seen as something complementary to man, as measurable matter and as cultivatable material, as manageable property. The agricultural-medical-technical revolution of the past two centuries has caused dramatic changes in natural energy and nutrient cycles. Man has played the role of despot, of absolute ruler who subjects nature to his own will and sees only its instrumental value. Although this seems to be an oldfashioned view, essential elements of it can often be found in the vivid optimism of technocrats with a blind trust in technology and for whom growth has no limits. The mere dominance of instrumental thinking (nature is there for man to be used and managed) is now perceived by many as one of the main roots of the environmental crisis.

As a result of the negative spin-off of the despotic attitude, more and more support of the view in which man sees himself as a partner of nature is developing.

Man as Partner

In this view man respects the intrinsic value of nature, existing side by side with man. In such a partnership, a dynamic relationship based on equality is developed. Interests and values of both partners are respected in order to develop jointly in a balanced way. Man interferes with nature in such a way that he does not harm its intrinsic value. He bases his actions on ecological knowledge. Both partners are expected to benefit from this mutual relationship. With this attitude man can play an essential role in order to improve the disturbed relationship which developed as a result of the absolute ruler attitude. The problem with this view is that it implicitly assumes that man knows how to improve the disturbed relationship, that man is willing to sacrifice and not apply technologies which appear economically attractive but environmentally unacceptable.

The content of this section is based on information presented in Zweers (1989) and Achterhuis (1990).

CONCLUSIONS:
FROM *HOMO ECONOMICUS* TO *HOMO ECOLOGICUS*

The acceptance of and respect for an intrinsic value of nature will result in an attitude which will make possible an acceptation of a new approach for the solution of environmental problems. Such an approach will consist of environmental policies of a radical nature and will demand a world-wide stabilization of the human population and a reduction of materialistic growth. It is an approach which cannot be realized through negative or positive sanctions alone, but which needs to be based upon a broad social acceptance. Education and information at all levels on the fragile biosphere of Mother Earth will be the crucial elements needed to accomplish this acceptance.

Many of the present day proposals to improve the deteriorating situation, including the Brundtland report, are aimed at sustainability and management of the environment (not nature). Their main message is: provide the necessities for now but do not risk the possibilities for survival of future generations. They do not speak about an earth which makes life worthwhile, but about a planet with opportunities for survival. Such ideas have been expressed since the 17th century (Achterhuis, 1990). They have never led to improvement of nature, on the contrary, the situation has never been as bad as today. It is very disappointing to have to conclude that the Brundtland report does not contain a single new idea which has not already been expressed before. Why then should its suggestions for improvement now suddenly be taken up and result in success? Why do all warnings about deterioration of nature since the mid 1650's not result in actions? How does one stop the man-caused pollution of nature which has already proceeded at full speed for more than 3 centuries? Do we wait till the physical limits of nature are reached?

The issue here is: is man able to establish and subsequently respect such limits. In my opinion he is far from knowing the latter. Serious studies of ecosystems were initiated some 200 years ago. Due to the great complexity of ecosystems, we are only aware of some of the relationships and feed-back mechanisms. We lack the most elementary understanding about the functioning of hundreds of thousands of the participants in ecosystems we lack the most elementary understanding. It is therefore hazardous and arrogant to manipulate such ecosystems as we do now. When we realize and acknowledge our poor understanding of nature, we should be able to accept our role as partner or participant instead of ruler, and act accordingly.

Finally, let me return to the role of the entomologist with regard to nature. The spin-off of his current activities to control crop pests and vectors of human diseases with chemical pesticides can be classified as negative to nature. *"...While the prospects for the future of insectkind is excellent, the prospect for the future of mankind is not at all that bright. Our environment is becoming a load of self-inflicted pollution..."(Odhiambo, 1977).*

Besides rejection of the use of conventional chemical pest control based on how we see nature, also new cost/benefit analyses taking indirect costs into account, will lead to a reappraisal of the value of environmentally safer control measures, particularly biological control. In order to achieve drastic changes, pest control should not be viewed as an isolated activity, but within an ecosystem approach. In such an approach, many more opportunities for long-term environ-mentally safe control methods may be achieved.

ACKNOWLEDGEMENTS

G.A. Pak, W. Takken and D.C. Thomas are thanked for thoroughly reading drafts of this paper and for important suggestions to improve the text.

REFERENCES

Achterhuis, H.J. (1990), Van Moeder Aarde tot Ruimteschip: Humanisme en Milieucrisis. Inaugurele Rede, Landbouwuniversiteit Wageningen, 40 pp.

Besemer, A.F.H. (1985), Veertig Jaar Volhardend Zoeken: Het "Ideale" Bestrijdingsmiddel Een Utopie? Afscheidscollege, Landbouwhogeschool Wageningen, 26 pp.

Bodenheimer, F.S. (1951), Insects as Human Food. Junk, The Hague, 352 pp.

DeFoliart, G.R. (1989), The Human Use of Insects as Food and as Animal Feed. Bulletin of the Entomological Society of America, 35, 1: 22-35.

Dover, M. and B. Croft (1984), Getting Tough: Public Policy and the Management of Pesticide Resistance. World Resources Institute Study 1, 80 pp.

Hogue, C.L. (1987), Cultural Entomology. Annual Review of Entomology, 32: 181-199.

Huffaker, C.B., F.J. Simmonds and J.E. Laing (1976), The Theoretical and Empirical Basis of Biological Control. In Theory and Practice of Biological Control (C.B. Huffaker and P.S. Messenger, eds.), pp. 42-80. Academic Press, New York.

Metcalf, R.L. (1980), Changing Role of Insecticides in Crop Protection. Annual Review of Entomology, 25: 219-256.

Nefyto (1990), Landbouw en chemische gewasbescherming in cijfers, 8 pp.

Odhiambo, T.R.O. (1977), Entomology and the Problems of the Tropical World. Proceedings XVth International Congress of Entomology, Washington D.C., August 1927, 1976, pp. 52-59.

Passmore, J. (1974), Man's Responsibility for Nature. Scribner, London, 214 pp.

Perkins, J.H. (1982), Insects, Experts, and the Insecticide Crisis. The Quest for New Management Strategies. Plenum Press, New York, 304 pp.

Pimentel, D., Krummel, J., Gallahan, D., Hough, J., Merrill, A., Schreiner, I., Vittum, P., Koziol, F., Back, E., Yen, D. and Fiance, S. (1978), Benefits and Costs of Pesticide use. Bioscience, 28, 72: 778-784.

Pimentel, D., Andow, D., Dyson-Hudson, R., Gallahan, D., Jacobson, S., Irish, M., Kroop,S., Moss, A., Schreiner, I., Shepard, M., Thompson, T., and Vinzant, B. (1980), Environmental and Social Costs of Pesticides: A Preliminary Assessment. Oikos, 34: 126-140.

Rodman, J. (1983), Four Forms of Ecological Consciousness Reconsidered. In Ethics and the Environment, D. Scherer and T. Attig (eds.), Prentice-Hall, Englewood Cliffs, 236 pp.

Smith, R.A., T.E. Mitler and C.N.Smith (1973), History of Entomology. Annual Reviews Inc., Palo Alto, California, 517 pp.

Southwood, T.R.E. (1977), Entomology and Mankind. Proceedings XVth International Congress of Entomology, Washington D.C., August 19-27, 1976, pp. 36-51.

USDA (1965), Losses in Agriculture. U.S. Department of Agriculture. Agricultural Handbook No. 291. Agricultural Research Service, U.S. Government Printing Office, Washington D.C.

van der Vaart, L. (1987), Indirekte effekten van bestrijdingsmiddelen, een aanzet tot kwantificering. Amsterdam, Wetenschapswinkel Vrije Universiteit, 187 pp.

van Lenteren, J.C. (1983), The potential of entomophagous parasites for pest control. Agriculture, Ecosystems and Environment, 10: 143-158.

van Lenteren, J.C. (1986), Parasitoids in the greenhouse: successes with seasonal inoculative release systems. In Insect Parasitoids (D.J. Greathead and J.K. Waage, eds.), pp. 341-374. Academic Press, London.

van Lenteren, J.C. (1987), Environmental manipulation advantageous to natural enemies of pests. In IPM Quo Vadis (V. Delucchi, ed.), pp. 123-163. Parasitis Symposium Book, Geneva.

van Lenteren, J.C. (1990), Implementation and commercialization of biological control in West Europe. Proceedings of Symposium 100 Years of Biological Control, A Century of Successes, 4-7 April 1989, McAllen, Texas, USA, in press.

van Lenteren, J.C. and Woets, J. (1988), Biological and integrated control in greenhouses. Annual Review of Entomology, 33: 239-269.

Vereijken, P.C., Edwards, C., El Titi, A., Fougeroux, A., and Way, M. (1986), Proceedings of Workshop on Integrated Farming, Bulletin I.O.B.C./W.P.R.S. 1986/IX/2, 34 pp.

Zadoks, J.C. ed. (1989), Development of Farming Systems: Evaluation of the five-year Period 1980-1984. Pudoc, Wageningen, 90 pp.

Zweers, W. (1989), Houdingen ten opzichte van de natuur: de aarde verdraagt haar heersers niet. Heidemijtijdschrift, 1989-3: 74-79.

World Commission on Environment and Development (1987), Our Common Future. Oxford and New York: Oxford University Press, 383 pp.

13.

MAN AS SELECTOR - A DARWINIAN BOOMERANG STRIKING THROUGH NATURAL SELECTION

LARS MUNCK
Carlsberg Research Laboratory
Gamle Carlsberg Vej 10, DK-2500 Valby, Denmark

*"It is an important principle
that in the process of selection
man almost invariably wishes
to go to an extreme point"*

Charles Darwin, 1868, Vol. II p. 226.

ABSTRACT

Since prehistoric times man has had the capability, stimulated by the diversity of nature, to select and utilize various items from his environment. At first man influenced only his immediate surroundings to improve his life and survival, but later he seriously intervened with the total global environment by exploiting the fossil energy resources. In this path, agriculture has constituted the basis of human development, securing a food store which enabled man to plan and to specialize in fields like administration, military defense, teaching and technology as well as in science, philosophy and arts.

In the light of the original work by Darwin this paper first discusses the modern plant and animal breeder as a selector driving his selection cycles, as a heuristic model of the basic human activity of experimental behaviour. This behaviour is also applicable for consumers in supermarkets and is now studied in social science as value theory. The development of human civilization based on agriculture is further looked upon as a gigantic breeding process involving not only plants and animals but also machinery and human concepts, such as law and money.

Our environmental dilemma is seen as caused by the unintentional side-effect of the selection force of the now more than 5 billion human beings selecting their means for survival and enjoying life on our common planet. The serious limitations of the human mind, including the mind of scientists, to overview and predict complex systems are elucidated by the model.

New strategies based on data systems and interdisciplinary thinking may bring us further in planning our future by learning from our mistakes. This will, however, not be effective if we do not realize in time the true consequences for man and environment of the concepts inherent in our culture, such as the monetary concept, many of which originate from the implementation of agriculture.

Key Words: Agriculture, Biology, Darwin, Economy, Environment, Natural Selection, Philosophy, Quality Control.

ON AGRICULTURE AND ENVIRONMENT

Up to the 18th century, our agricultural and craft based society lived by harvesting renewable crop and forest resources. Then something extraordinary happened: the emergence of industries independent of agricultural land which exploited the great stores of fossil hydrocarbons for energy and chemical synthesis. To the grain, food and timber stores could thus be added the much greater reserve of organic material stockpiled by nature over geologic time. However, the agricultural experience was instrumental in the exploitation of these resources, causing major changes in human thinking and living (see essays by Steensberg, 1986, 1989). Our change of approach is still in flux, and current concerns with the environment reflect our unease with our own adaptive process and the values we employ to rationalize it.

This essay is intended as a conceptual discussion of our environmental impact from an agricultural perspective. The conceptual framework of this essay is derived in part from a new science now slowly emerging, originating from a succession of physicists, including Bohr and Einstein (see books by Bohm (1980) and Prigogine and Stengers (1984)). This paper is not a Darwinistic exercise in the evolutionary sense Darwinism is thought of today. Instead, I will demonstrate how man not only manipulates plants, animals and environment, but also how he manipulates concepts such as those originally derived from Darwin. The result of this ruthless manipulation, even distortion, is a paradigm of the evolution of modern society fundamental to our present conflict with environment.

ON BREEDING AND SELECTION

My Professional Connection to the Environmental Concept

At the beginning of this century, biological chemists demonstrated a clear connection between the amino acid composition of different plants and the nutritional requirements of animals and man which ate them (Munck, 1990). Of the approximately 20 amino acids which comprise our proteins, at least 8 are essential to growth in higher animals. In diets based on cereals, lysine is the first limiting amino acid for animal and human growth. During the 1960's, malnutrition and its remedy became a major issue in the public, and many people had the impression that the welfare of mankind was entirely dependent upon the quality of protein we ingested.

Much of my scientific work for the past 25 years has been spent searching for and improving the nutritional quality of cereal grains (Munck, 1972). Working with colleagues in Sweden and Denmark, we isolated the first high-lysine barley using a screening method we developed to measure lysine in natural and mutant barley lines. Genetic analysis then showed that a few genes controlled improved grain amino acid balance, one of which produced a grain with a nutritional value close to that of milk. This was due to a change in the levels of different barley grain proteins.

Unfortunately, the yield of grain from these mutants was, due to small, shrivelled kernels, too low to be compensated for by their improved protein quality. We therefore started a breeding programme to see, by crossing and selection, if we could find a genetic background for the mutant gene giving higher yield. Luckily, the original mutant kernels had a large germ. Selection for this easily identified trait was used to substitute for chemical analyses, greatly speeding up the breeding program. High-lysine lines were developed by 1990 (Munck, 1990) with improved seed quality and with competitive yield, mainly due to an increase in the number of seeds produced per area unit. However, the protein issue had changed by then. Although important, protein was no longer seen as the limiting nutritional factor for much of the world's population. There was no "protein crisis" in the developing countries but problems of purchasing power and distribution. In Denmark and the EEC, feed (soy) protein was imported without duties, making it cheap compared to locally produced cereal protein. This made our improved barley less attractive, even though utilization of the grain could have saved the pig breeders 30-70 per cent of their need for protein imports if trade had been otherwise regulated. However, another reason for the problems encountered in marketing a high-lysine barley, even with competitive yield, was that the feed industry was not equipped to handle high-lysine barley with regard to lysine control and separate storage. The pig production industry was not geared to utilize it optimally because it involved both a change in the feeding regimes and a change in emphasis of production from bacon to larger solid meat cuts.

Nonetheless, environmentalists became interested when we realized that the new barley, when used as a pig feed, could reduce urine and faecal nitrogen excretion by up to 20 per cent. This is clearly a positive environmental effect in a small country producing about 17 million pigs per year!

Interestingly, the high-lysine genes were also found to enhance the synthesis in grain of a number of proteins having specific antimicrobial effects. The effect of the overexpression of these proteins is therefore of scientific interest as a tool for studying pathogen resistance in plants. Thus, spin-offs from our original research, which nobody thought about at the beginning of the project, were positive from an environmental point of view.

In conclusion, the primary breeding hypothesis launching a high-lysine barley was found to involve unexpected options and drawbacks on several levels, showing all signs of the fact that the society was not adapted to the high-lysine concept.

An Approach to a Working Hypothesis Involving Environment

Taking my experiences as an example, we can see that man's selection process (Figure 1) starts with a primary selection hypothesis in the global area. It is based on experiences of Surveying (I) (inventories) creating an attitude towards, the barley plant and its characteristics of general importance such as yield, seed quality and disease resistance, together with quality criteria such as pig yield.

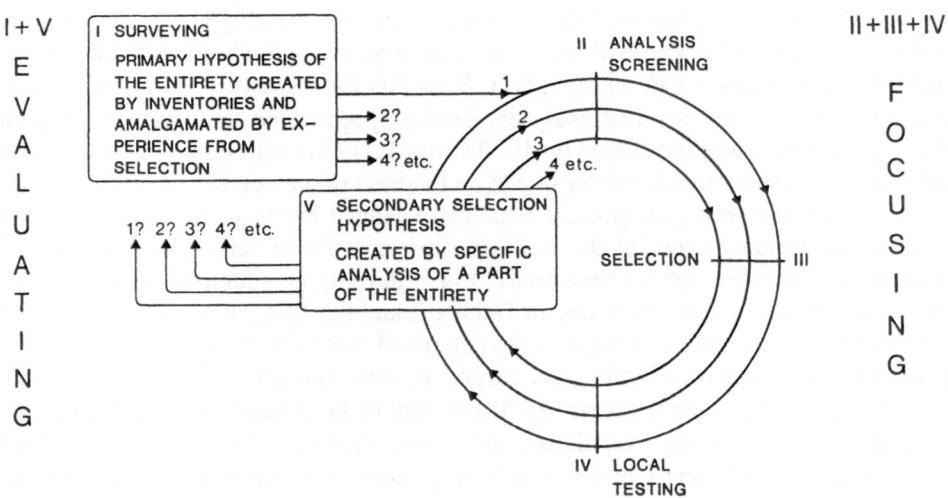

Figure 1. The selection cycle.

In the next Focusing steps (II-IV), a breeder, aided by science, analyses in detail an empirically set goal (e.g. better pig growth on barley) and finds that grain lysine is a limiting factor. The breeder then develops a screening method for this criterion, and applies it. The provisional and general selection criterion, in this case our mundane problem of pig growth is now exchanged for a more specific, secondary one - lysine (V). After analyzing the barley strains for lysine (II) the breeder selects the deviating strains (III) and tests them in the field for yield and quality (IV). After several years and numerous breeding and selection cycles, the breeder has a multifaceted, total experience of the dynamics of selection with regard to lysine. The efficiency of man as selector depends upon the following conditions which drive and limit the selection process:

1.   **The abundance of variation** - genetic variation is needed for selection progress (in breeding)

2.   **The nature of the selection criteria** - visual (field) characters (such as germ size) are easier to work with than characters (such as lysine) which have to be analyzed

3.   **The number of selection criteria** - it is more complicated to select for five characteristics than for two

4.   **The efficiency in the testing and evaluation process** - handwritten notes are harder accessible than computer data

5.   **The number of selection cycles** - is related to experience

6.   **The communication with other selectors** - exchange of (plant) stocks and selection methods speed results

7.   **Resources** - money, skilled labour, etc.

8.   **The ability of evaluation** - to keep a dynamic exchange of views between the general primary hypothesis (I) and the specific secondary selection hypothesis (V) to combine synthesis with analysis.

This kind of selection process, manipulating hidden characters through visible, underlies much of human behaviour. In our example, the behaviour of a breeder is a "shadow play" on at least two levels: the barley genotype and the needs of the downstream production chain. In this play breeders select phenotypic characters, such as yield and lysine content, to sort out genotypes adapting them to the needs of the market. In much the same way, a consumer moves around a supermarket to select foods by their visible characters within a limited budget to fulfil certain needs. Some of the visible and hidden characters (values) underlying food selection are listed in Table 1.

Table 1.
Consumer's selection criteria/values. (Munck 1972, p. 118)

Form	)	)
	)	)
Price	) evaluated	)
Prestige	) before	) apparent
	) purchasing	) visible explicit
Colour	) (selecting)	) characters
	)	) (values)
Convenience	)	) - secondary
		) evaluation
		) (V, Fig. 1)
Cooking quality	)	)
	) evaluated	)
Smell	) at cooking	)
	) and eating	)
Taste (immediate	)	)
well-being)	)	
Growth of	)	)
children	)	) implicit, hidden
	) evaluated	) characters
Maintenance		
of adults	) as long term	) (values)
	) viability	) - primary
Health	)	) evaluation
	)	) (I, Fig. 1)

The success of a plant breeder or a consumer depends upon the selection of visible characters commanding their primary attention, but which are later metamorphosed on the hidden level as income from a marketed variety for the breeder or good health for the consumer. After operating the selection process through numerous cycles, experience about the nature of the hidden values accumulates and can be visualized in subsequent cycles as primary evaluation. Obviously there should be some coherence between the visible and hidden selection criteria for the successful life of a plant breeder and a consumer. Over time, then, previously hidden values become visible and open for discussion and criticism, as demonstrated in the case of high-lysine barley previously outlined.

Today, the dynamics of the production chain from breeder to consumer depends upon the apparent value of money. Older peasant societies, which existed without the money concept, were hard to live in. Individual members of such smaller societies, however, had an immediate awareness of their entire production chain (a primary hypothesis). They knew intimately that badly dried, mouldy grain caused sickness in animals and human beings. In our modern society, money, originally created as a representation of surplus, has impelled a specialization into various grades of producers and consumers. As a result, we no longer rely on our local, empirical experience. Money, exchanged with goods and services between manufacturers, farmers, wholesalers, retailers and consumers, and organized by bureaucrats, technocrats and politicians has become the foremost visible selection character and has introduced the needs for an objective quality control in each transaction. Without continuously visualizing the hidden values by quality definition and pricing, they will all succumb in the production chain.

ON DARWIN AND ENVIRONMENT

Man's achievements in plant and animal husbandry inspired Darwin (1868) to develop a generalization in the form of the concept of "natural selection" (1859). My experience as a plant breeder and scientist has led me to invert the Darwinian perspective of natural selection. This inversion entails contemplating the role of man as a selector and "breeder" of not only plants, animals and earth, but of concepts in his mind. Thus, "*Selection and exploitation by man of the world resources are now being regarded as the most singular and important component of natural selection and survival with respect to all living organisms*" (Munck, 1972, p. 120).

The essence of the environmental problem is that while expressing our individuality through buying visually identified objects, we are unaware of the dynamic, hidden effects of our selections. We then are surprised, even insulted, when the environment from which we selected, changes, or even strikes back. This environmental reaction resulting from the side effects of human selection is a **Darwinian boomerang**, in the end striking man and other species through natural selection. Its force largely originates from the mass action of individuals, as selectors of objects and resources. Therefore, the individual selector is the ultimate principle to address - all other "organizations" or "people" are abstractions which shadow this principle.

The impact of Darwin's work "The Origin of Species" has been tempered, fragmented, distorted and exploited by a wide range of "dealers" in human thought. As a result, we still have not drawn realistic conclusions required for our survival (Gould, 1977; Waters, 1986).

For example, the word "evolution", now irrevocably tied to Darwin, was coined much earlier than "The Origin", and was used only once by Darwin himself. He used the "neutral" expression "*descent with modification*", instructing himself in his notebook "*never say higher or lower*" (Gould, 1977). He was aware that natural selection produced a full range through descent - both human beings as well as parasites which emerged as simple, degenerated sacks of tissue. The positively charged term *evolution*, a secondary hypothesis, came to substitute Darwin's "*descent with modification*", even against his will. Such an adoption was favoured as it legitimized the positive trends in the mushrooming industrial/colonial society of that time. Evolution in and of science, technology and social sciences obviously was touted by the extreme market selectors. In my own field of work, besides the protein issue, I have witnessed similar rationalizations adopting the "green revolution" from the breeders in the 1960's, and the rise of biotechnology in industry from the scientists during the 1980's.

While studying the explicit world in the form of the adaptation of wild life to the island environments of the Galapagos, and by collecting data regarding the domestication of plants and animals, Darwin realized an implicit world as real as the explicit which he expressed in the form of the concept of natural selection. Natural selection embodied all the different forces acting upon inherent genetic variation, modelling it to fit special environments through a descent with modification. Although he (Darwin, 1871, p. 73) acknowledged human inventiveness "... *These several inventions by which man in the rudest state has become so preeminent are the direct results of his powers of observation, memory, curiosity, imagination and reason*". He (Darwin, 1859, p. 2) could not, however, see the boomerang of the human selector returning in the form of natural selection "*From a remote period in all parts of the world man has subjected many animals and plants to domestication or culture. Man has no power of altering the absolute conditions of life; he can not change the climate of any country, he adds no new element to the soil but he can remove an animal or plant from one climate or soil to another ...*".

Furthermore, Darwin (1859, p. 2) could not forecast the consequence of man's future insights in genetics, "*If organic beings had not possessed an inherent tendency to vary, man could have done nothing. He unintentionally exposes his animals and plants to different conditions of life and variability supervenes which he can not even prevent or check*". Thus Darwin discovered and formulated the hidden scene of natural selection as a primary hypothesis, which incorporates, without any formal difficulties, aspects he was unaware of at that time, the unintentional environmental boomerang caused by man as selector and man's ability to intentionally manipulate genomes and the environment. I cannot help wondering whether Darwin, an excellent observer, would have predicted the potential force on environment of man as selector if he had inspected the industrial backyards near his home with the same care as he spent studying finches in South America. Indeed, the current environmental issue might have been expressed and tackled differently if Darwin and his followers had succeeded in or **had been allowed to** open this door to the primary evaluation area, keeping the intellectual climate of his time in mind.

ON THE RELATIVE IMPORTANCE OF GENES AND ENVIRONMENT

Secondary Evaluation

On a visit to the Institute of Animal Genetics, Novosibirsk, the director, Dr. Belyaev, demonstrated the outcome of an experiment selecting foxes for aggressiveness or for domesticated behaviour. Dr. Belyaev carried out this experiment for more than 20 years, in which he selected foxes at 2-4 months for breeding separately in aggressive and domestic populations (Belyaev, 1978). The aggressive foxes displayed excellent, shiny furs, and loudly protested when we cautiously approached their cages. The foxes from the population showing domesticated behaviour were reared in an old house with open doors surrounded by a fenced area. When I stepped into this enclosure, several foxes jumped up upon my shoulders to lick my face, showing an overwhelming social affection and friendliness. The quality of their furs, however, was more typical of poor dogs than of wild foxes or of the foxes from the aggressive population. Dr. Belyaev had further studied the heritability and physiology of the aggressive/domesticated behaviour traits and shown that these traits were dependent on genes of a putatively regulative character. He demonstrated that the selection had affected the hormonal balance of the animals, which putatively explained why the domesticated foxes lost their seasonal oestrous cycle, as well as their aggressiveness and high-quality fur.

Let us now **cautiously** transfer the essence of this experiment on selection to man himself. Assuming for a moment that aggressiveness/domesticated behaviour also has a genetic/hormonal basis in man, it would appear that he has domesticated himself by eliminating aggressive human individuals, here defined as a part of natural selection acting on the survival of man. There are indications that the human brain and muscular strength have remained the same over the last 20.000 years, in part because we have manipulated our environment rather than vice versa. However, selection pressure from society toward domesticated behaviour involving the hormonal system could have been in operation later in the history of man. If this had been the case, the result of man's "domestication of man" has wrought a subtle balance between aggressiveness and social principle - our headache and inspiration.

Primary Evaluation

With Dr. Belyaev's demonstration in mind, I fully appreciate arguments on the importance of the genetic basis of behaviour put forward by Konrad Lorenz in the 1974-edition of Darwin's work "On the Emotions in Man and Animals". In his preface to this cornerstone of the science of ethology, Lorenz said "*... behaviour patterns are just as reliable and conservative characters as are the forms of bones, teeth and other bodily structures*". On the other hand we are all aware of the impact of our environmental conditioning during childhood, emphasized in numerous treatises in social science.

Both of these observations are valid as secondary selection results (Figure 1). However, we often make a fundamental mistake in our thinking when we put them **against** each other to attempt a synthesis to improve our primary description of the forces, by which we evolve the hidden scene behind the individual or group identities. This is because genes and environment do not affect us as separate factors, but as amalgamated latent factors. Only in extreme cases, such as a haemophilia or a lung cancer, are we able to blame their condition on genes or the environment. Indeed, we have to accept and live with the uncertainty in the dynamics of the hidden scene on formal grounds such as that exposed in Heisenberg's uncertainty principle: just as we can never know the exact position and velocity of an electron, we can never (although on other grounds) in each individual case know how much of the behaviour of an individual is determined by genes or by environment.

Darwin understood the importance of constantly tempering his own selected extremes of human thought. In "The Descent of Man" (1871, Part II, p. 404) he wrote "*Important as the struggle for existence has been and even still is, yet as far as the highest part of man's nature is concerned there are other agencies more important. For the moral qualities are advanced, either directly or indirectly, much more through the effects of habit, the reasoning powers, instruction, religion etc. than through natural selection; though to this latter agency the social instincts, which afforded the basis for the development of the moral sense, may be safely attributed*". Given the information from our secondary evaluation, we can thus tentatively interpret Darwin's words in a primary hypothesis, concluding that: **the domestication of man through natural selection afforded the necessary genetic base for a social principle which is advanced directly or indirectly much more through the effects of his culture and physical environment than through genetics.**

Thus, when modelling such a hidden scene by selection, we have to weigh explicit secondary qualities in our implicate primary evaluation and bear in mind and respect the uncertainties conceptually and analytically surrounding our methods. Hence, in the primary evaluation stage, **never say either - or, say both, not just together but amalgamated with all other factors in the entirety.** Remember that polarizing, by focusing on the extremes (see the citation from Darwin under the title of this paper), is just a method for gathering information (a secondary selection hypothesis) throughout the selection cycle. Such polarizing is not necessarily transferable from local to global conceptual areas. Thus, while it is a good idea to recognize and test extremes, focusing on the local analytical area, we can not afford the same luxury when we test a primary evaluation in the global area - too much is at stake then.

It seems that much discussion about the importance of genes versus environment in natural and social sciences after Darwin (including large parts of social Darwinism and behaviourism) would have been unnecessary if his followers had grasped this holistic methodology.

THE ENVIRONMENTAL ISSUE AS A CONCEPTUAL CRISIS

More than 5 billion active human selectors are now keeping our complex production chains busy. Side-effects from producing and consuming foods, cars, refrigerators, and weapons are now major parts of natural selection, the mass action of which is ironically called "environmental effects". In the short term, these selections strike back at us occasionally, tempering the quality of our lives. In the long run they may upset the entire biosphere. Even though we never will be able to completely overview or control our global situation, this kind of control is, however, what the public today expects from science and technology - rather unrealistically.

The concept "man as selector" is, I think, useful in searching for a common denominator in the multifaceted phenomenon "Homo sapiens". This heuristic concept, and the selection cycle as a model approach, includes at least the eight limitations earlier described. Additionally, it includes the fundamental conceptual difference in strategy between on one side experimentation in localized conceptual areas of the selection cycle and on the other side evaluations in global areas required for the implementation of data or ideas. Such a heuristic approach is also necessary to understand and communicate with the extremist forces among the public that seek to scrap our technological tools because of their involvement in environmental deterioration. This group of people is caught in just the same secondary selection treadmill in the local conceptual area as their counterparts - the positivistic evolutionists. Instead, we have first to temper our utilization of the many fruits of science and technology, improving our evaluations in primary areas with practical and ethical arguments. In particular, we need to start with provisional, primary selection hypotheses other than monetary income, because money belongs to evaluation in the secondary area. Given the forceful tools of modern technology, it should now be considered unethical to be solely pragmatic in management, confusing monetary earnings with a true primary hypothesis of activity.

To communicate with the market of individual selectors, we need to exchange our present paradigm or primary hypothesis - evolution in monetary economy through consumption - into that of **a structural development within a limited framework, on ethical grounds as reflected in monetary tools, emphasizing population control, recycling, sustainability and quality definition and control**. We must, however, find new and unbureaucratic ways of realizing this control. For example, doubts regarding the feasibility to measure quality in food products have created both the biodynamic movement and the ISO 9000 industrial standard, favouring certification of production to avoid too elaborate, costly and late analysis of the end-products. Likewise, the concept "Total Quality Management and Control" (Ishikawa, 1985), which is behind the success of much of Japanese industry, emphasizes a holistic approach to production, including its environmental and ethical aspects.

There is clearly a need for more hard thinking, illustrative examples and application successes. To support the holistic concept, it is not sufficient to cry out for *holism* in *the media* just as was and still is done for *evolution, competition in the free market* and *economy* since the 1860's.

There are now mathematical languages which can help to define logical thinking on latent factors and implicate order. For example, using partial least square (PLS) and partial component analysis (PCA) (Martens and Næs, 1989), process control data sets can be viewed as multidimensional, dynamic, non-linear representations of the brewing process from barley to beer. As far as I can judge, these multivariate tools connect to the vector algebra concepts of implicate order in Bohm's quantum physics (Bohm, 1980). Indeed, reading Bohm, and contemplating my meetings with contemporary breeders and brewers as well as subsistence farmers, I am convinced that the human brain deals primarily with whole concepts rather than deriving concepts by putting together fragments, or isolated data, which appears to be a secondary thought process.

Apparently, we are slowly learning how to adopt screening and computer capabilities to make a primary evaluation of the multi-factorial data sets which are reality in science. We have far to go in communicating the conclusions to the market of selectors as a convincing paradigm, because of the tendency of the media to emphasize extreme, concrete events and ideas from the local, secondary selection area instead of a balanced primary evaluation, including an explanation of the degree and nature of the uncertainty which is involved.

Indeed the paradigm of "the new science" (Bohm, 1980; Prigogine and Stengers, 1984) suggests that it is neither possible nor practical to count on completely controlled situations and systems, such as those emphasized in reductionistic approaches in classical science. Since Heisenberg, physicists began to doubt their virgin belief in potential, unlimited manipulation by man. They have slowly started to realize that they are actually part of their own experiments! In the same way, biologists and social scientists could accept other kinds of uncertainty principles e.g. in the relative importance of genes and environment to an individual, avoiding the extreme conclusions of social Darwinism and behaviourism in the final, provisional primary evaluation. However, this ongoing change of paradigm in science has hardly reached company leaders, politicians, governments and media yet. They are definitely having unrealistic expectations of what science could judge and provide in a primary evaluation, which creates an intellectual climate no less distorting and repressive than that of the time of Darwin.

Therefore, to advance conceptually in the primary evaluation area, we need to realize new interdisciplinary incentives, to understand the true consequences of concepts inherent in our cultures. Many of these concepts are derived from agricultural man who selected and experimented in local realms of cultivation and production. They include the ownership of land and buildings, capital in the form of a food stores or money, and specialization, such as the military force originally established to guard capital stores. These concepts are fixed in our collective experience by literary and mathematical languages (Steensberg, 1986, 1989).

As individual human selectors with different professions and tools, we have different visible and hidden worlds and have empirically found different relations between them. Thus, we have often deduced different secondary selection hypotheses from essentially the same phenomenon, often without realizing it. An instructive, conceptual example is here the multifold manifestations on the phenotype level of a high-lysine barley gene studied with different methods by several specialists to find out which phenomena are primary and secondary (Munck, 1990). This complex constitutes "a cage of covariance" which the geneticists call pleiotrophy, a hidden scene only to be opened up by finding the gene sequence of the high-lysine gene in order to study its implications in the living system. Therefore, by cutting different planes in the "Gordian knot" and comparing the incisions, we can undoubtedly collectively obtain higher levels of holistic understanding, but it will definitely have analytic and conceptual limits.

The "new science" will give birth to a new thinking in monetary economy and again identify the humanistic disciplines in their own right, taking care of primary evaluation and revalidating their present deviation towards a reductionistic approach copied from the classical natural science.

A major problem in conceptually communicating the new science is language. Part of this is because the evaluation of uncertainty, latent in the new science, is anchored in concepts of the unknown, untouched and isolated in the classical, deterministic natural science, which still dominates the scene, leading to a loss of spirit of the world for the individual. Connecting to our humanistic heritage, indeed more enlightened reviews on approaches to the "hidden scene" throughout our history, including those of Plato, Adam Smith and, among all, Darwin, would help us in approaching the environmental issue, regaining our respect and fascination for the complexity of nature.

The stages of the selection cycle of individual, experimental man concerns his specific activity (breeding, farming, etc.), his stage of life, and the historical collective development of his culture. But they are interpreted and translated in his language! Language structure therefore plays a major role in the fragmentation of our thoughts, containing both static (nouns) and dynamic (verbs) elements which are leading and tempering our creativity. In order to ameliorate dynamic, holistic thinking, Bohm (1980) has aimed at focusing on giving verbs a stronger role than nouns. His rheomode meta-language (rheo from a Greek verb meaning flow) is based on present English. In Table 2 I have tried to relate the vital verbs of Bohm's rheomode language with principal sequences of the selection cycle (from Figure 1), also including interdisciplinary connections to other areas.

Thus vidate (to overview) belongs to sequence I Surveying, dividate (to divide) to the sequences II-IV Focusing and Factate (to make fact) is essential for the sequences V and I Evaluating. In further selection cycles, one is thus revidating, redividating and refactating to differentiate and select objects and processes from those which are irrevidant, irredividant and irrefactant.

In natural (classical) and social sciences, empiricism and sociology incorporate much of the strategy of surveying. Reductionism/determinism and psychology incorporates focusing, while technology/ecology and politology include evaluating.

Table 2
Interdisciplinary aspects of man as selector.

AREAS	STAGES IN THE SELECTION CYCLE[1]		
	I. Surveying Creation of a primary hypothesis on the entirety based on descriptions of ratio and reason of the world	II-IV. Focusing Characterization, selection and local testing of objects and processes for partite characters with specially devised methods	V+I. Evaluating Dynamic realization of the selected objects and processes in a secondary (V) hypothesis amalgamating it with the primary (I) one
Linguistics[2] First selection cycle Further selection cycles	Vidate Ordinate Revidate-Irrevidant Reordinate-Irrordinant	Dividate Levate Redividate-Irredividant Relevate-Irrelevant	Factate Constatate Refactate-Irrefactant Reconstatate-Irreconstatant
Natural sciences	Empiricism	Reductionism Determinism	Technology Ecology
Social science	Sociology	Psychology	Politology
Letters[3]	Epics Woman Nature	Lyrics, poetry[2] Man The ego	Drama Marriage The ego as selector in nature
Religion and outlook on life	The Gods are in nature in which man is living	God is near to man Man rules nature	Matter contemplates itself and aims at tempering itself through man

1) The stages of the selection cycle of man as outlined in Figure 1 in this paper.
2) Bohm (1980).
3) Nielsen (1990).

According to Nielsen (1990), Hegel noted that in the historical development of literature, epics (Homer) were primary (surveying), lyrics (the Greek poets) were secondary (focusing), and drama (the Greek dramatists) was tertiary (evaluating). Bohm (1980) in harmony with this analysis characterized the classic, reductionistic scientist as a poet. In literature according to Nielsen (1990) woman was considered epic, man was lyric, and marriage drama. In a similar way, religions began as pantheisms (God in nature) which corresponds to surveying. In the focusing stage there is only one God who is near to man, and man rules nature. Finally, in the evaluating stage, matter contemplates itself through man. **Thus the environmental issue can be defined as a drama of the ego of man and woman as selectors in nature, of how they, by increased consciousness of their connection to natural selection, could control their activities so that matter through** them **could temper itself for a moment.**

EPILOGUE

It is important that we learn to enjoy life in and through activities. To do this, we should aim for a balance with nature in our selection of resources, which we never will fully obtain but which we could improve, by making new innovations in human communication and cooperation. The activity itself is our reward, the objects we are selecting along the roadside are only signs and symbols of the fact that we have lived, essentially a kind of surplus.

It is consoling that historical examples exist in e.g. old Polynesia (Clarke, 1990) which demonstrate how man could learn from agriculturally induced environmental catastrophes and establish sustainable development by population control and a balanced utilization of nature. To achieve this on a global basis we have first to realize the true consequences of the concepts inherent in our culture. History teaches us that our conceptual heritage, including that obtained from Darwin as discussed here, is under constant selection and modification by "man as selector". We are writing in sand, on which wind is blowing.

Let us, therefore, with an open mind aim to fundamentally reconstruct this flux of conceptualization, largely originating from the agricultural experience, and its implications on man and environment. **By knowing the bias of our compass we could correct our course!**

ACKNOWLEDGEMENTS

I am grateful to Dr. John Mundy and Mrs. Kirsten Kirkegaard who have assisted in completing the manuscript.

REFERENCES

Belyaev, D.K. (1978), Destabilizing selection as a factor in domestication. The Wilhelmine E. Key 1978 Invitational Lecture. J. Heredity (1979), 70: 301-308.

Bohm, D. (1980), Wholeness and the Implicate Order. Ark Paperbacks, London, England, 224 pp.

Clarke, W.C. (1990), Learning from the past: Traditional knowledge and sustainable developments. The Contemporary Pacific, 2: 233-253.

Darwin, C. (1859), On the Origin of Species by Means of Natural Selection. Watts and Co., London, England, reprint of 1st ed. 1872.

Darwin, C. (1868), The Variation of Animals and Plants under Domestication. 2nd ed. 1890, Vol. I-II. J. Murray, London, England.

Darwin, C. (1871), The Descent of Man, Vol. I-II. J. Murray, London, England.

Darwin, C. (1874), The Expression of the Emotions in Man and Animals. The University of Chicago Press, Chicago, USA, 1974.

Gould, S.J. (1977), Ever Since Darwin - Reflections in Natural History W.W. Norton & Company, Inc., New York, USA, 285 pp.

Ishikawa, K. (1985), What is Total Quality control? The Japanese Way. Prentice-Hall, Inc., Englewood Cliffs, N.J., England.

Martens, H. and Næs, T. (1989), Multivariate Calibration. John Wiley & Sons, Chichester, England.

Munck, L. (1972), Improvement of nutritional value in cereal. Hereditas, 72: 1-128.

Munck, L. (1990), The case of high-lysine barley breeding. (Manus. in prep., see Editor's note on p. 227).

Nielsen, E.A. (1990), Livsbilleder - om digtningens udtryksformer. Danmarks Radio, Copenhagen, Denmark.

Prigogine, I. and Stengers, I. (1984), Order out of Chaos - Man's New Dialogue with Nature. Fontana Paperbacks, London, England, 349 pp.

Steensberg, A. (1986), Man - The Manipulator. The Royal Danish Society of Science and Letters, Copenhagen, Denmark, 200 pp.

Steensberg, A. (1989), Hard Grains, Irrigation, Numerals and Script in the Rise of Civilization. The Royal Danish Society of Science and Letters, Copenhagen, Denmark, 147 pp.

Waters, C.K. (1986), Natural selection without survival of the fittest.
Biology and Philosophy, 1: 207-225.

Editor's Note
The author has recently worked more specifically on some technical and biological aspects related to agriculture and environment. For further information interested readers may wish to apply directly to the author (address above). This additional work is planned for publication as follows:

Munck, L., On the utilization of the renewable resources. In Plant Breeding: Principles and Prospects. (M.D. Hayward, N.O. Bosemark and I. Romagosa eds.). Chapman and Hall, London, England.

Munck, L., The case of high-lysine barley breeding. In Barley: Genetics, Molecular Biology and Biotechnology. (P.R. Shewry ed.). CAB International, Wallingford, England.

14.

PHILOSOPHY AND THE ENVIRONMENTAL CRISIS

JOHN H. FIELDER
Philosophy Department
Villanova University
Villanova, PA 19085, USA

ABSTRACT

Environmental problems are partly the result of conceptions of our relationship to nature. I present a critique of selected beliefs about our relationship to nature that have contributed to the environmental crisis and some others that I believe will help create a more ethically acceptable conception.

Key Words: *Environment, Ethics, Future, Nature, Patriarchy, Philosophy, Rights, Self-interest.*

INTRODUCTION

If you ask the question "What does philosophy have to do with the environmental crisis?" there are two possible answers; "not much" and "everything." Both are correct, although not in the same way. "Not much" aptly characterizes the role of professional philosophers in environmental decision making. Philosophers who analyze the beliefs and values underlying those decisions have little influence outside the university and not much more within. President Bush does not have a Council of Philosophical Advisors, there is no National Philosophy Council, nor can we expect to see a cabinet position dealing with philosophical matters.

On the other hand, fundamental beliefs about knowledge, nature, value, life, and morality powerfully shape thought and action. It is one's philosophy that defines what counts as a problem and what will be regarded as an acceptable solution. It is in this sense that philosophy - what people believe, particularly about the basic questions of life - has everything to do with the environmental crisis and our response to it.

Public discussion of environmental issues tends to focus on the scientific, engineering and economic problems associated with meeting environmental challenges. The spot-light is occupied by questions about how our present practices are threatening the biological conditions for life, what technical changes are possible, and how much it will cost. For example, in the United States the government has declared the northern spotted owl to be an endangered species, and this creates conflicts with the logging industry whose activities are destroying the owls' habitat. In the media the issue is presented as owls vs. jobs, with the usual "realists" favoring jobs and environmentalists defending the owl. Economists calculate the jobs-per-owl ratio. No one seems able to move the discussion to more fundamental questions about our dilemma: How do we conceive our relationship with the natural world? How has that conception contributed to the environmental crisis? Is there a better way to understand our relationship to nature?

The environmental crisis is forcing us to examine some of our basic assumptions about our relationship to nature. It is essential that these philosophical issues become part of the larger public debate on the environment, so that our technical and economic options are also seen as choices that reflect certain beliefs and values in a philosophy of life.

The Environmental Crisis

It is hardly necessary to list the unprecedented environmental concerns that face us in the latter part of the twentieth century. Oceans, rivers, lakes, air, land, plants, animals and our own bodies are laced with the toxic by-products of industrial civilization. Even more frightening than this is the prospect of global warming. Bill McKibben, in The End of Nature, (1989) points out that although the sky seems as boundless as the seas and the land, in fact it is remarkably finite. Seven miles on the surface of the earth is a very short distance today; many workers commute more than this distance to work. But seven miles vertically takes you to the edge of the atmosphere, right on the boundary of the ozone layer. The sky looks boundless, but the atmosphere is extremely limited, a thin skin of gas enclosing the earth.

A hundred and fifty years of industrial growth has poured enormous amounts of carbon dioxide and other gases into the atmosphere that surrounds us. As the composition of the atmosphere changes, heat does not escape as easily and the average temperature of the earth rises. This will result in significant - some say catastrophic - climatic changes that will dwarf our current worries about PCBs and nuclear waste (Golub and Brus, 1990).

A significant rise in the earth's temperature will pose the severest challenges to our ability to provide the basic necessities to the world's population and prevent widespread political instability and conflict. Most alarming is the fact that global warming is directly tied to our industrial practices, particularly the combustion of fossil fuels. The changes that will be required are therefore extensive and fundamental. They will require dramatic political, technical, and personal responses. Earth Day 1990 brought forth a new wave of publications dealing with environmental issues, most exhorting us to begin immediate action to forestall catastrophe.

The Role of Philosophy

Any significant response to the environmental crisis will require changes in living patterns brought about through political action. Neither will occur on the scale needed unless there is an appropriate change in how people think about themselves in relation to the natural world. It is not enough to simply point out the disastrous consequences of our present way of life; we must examine the ideas that led us to this situation and articulate a new vision of that relationship. Daniel Callaghan (1989) was writing about the health care system, but his words also fit the environmental crisis, for as long as the old values that brought us to this point *"pervade our culture and define our expectations, ...no amount of bureaucratic manipulation or procedural [changes] will bring meaningful reform...Such measures, without a change of values, can provide no more than temporary, symptomatic relief."* Similarly, Aldo Leopold, (1949) one of the seminal thinkers in environmental ethics, noted that *"No important change in ethics was ever accomplished without an internal change in our intellectual emphasis, loyalties, affections, and convictions."* The task of constructing strategies to deal with the environmental crisis must include new philosophical conceptions which will provide a better picture of our relationship to the natural world.

Søren Kierkegaard (1846) saw this clearly and expressed it as the distinction between objective and subjective truth. An objective truth is one that is true for any observer, while a subjective truth is one that is accepted as a basis for action by a particular person. A subjective truth is one that I care about, that I have incorporated into my thinking and acting. All the objective truths about what we are doing to the environment and how it can be changed will have no effect unless individuals appropriate them and see them as real and meaningful for them. That is why any program or analysis or call to arms must have a philosophy that criticizes our present beliefs as well as develops a new vision of our relationship to nature. It must build on what is already valued and show how this can be extended in a new way that is intelligible, desirable and practical. What philosophy can provide is a critical analysis of our present view of our relationship to the natural world and reasoned conceptions of alternatives. This means locating the environmental crisis within the framework of beliefs and values that helped create it and developing new conceptions that are more suitable to our present circumstances.

A complete review of the writings on this topic is impossible here; indeed they could not even be listed in the space of this paper. An excellent recent review of these topics, which I have relied on extensively in the preparation of this paper is Nash (1989). What follows is a selection from current philosophical literature of some beliefs that have contributed to the environmental crisis and some others that have a role in developing a better conception of our relationship to nature. For the sake of symmetry I have chosen three of each. The villains are Christianity, patriarchy and the industrial concept of the natural world as merely a commodity; their counterparts are self-interest, the needs of future generations, and respect for nature. My hope is that this approach can contribute to the development of new policies in which technical and philosophical issues are closer together in our thinking about environmental problems.

I

Nature as a Commodity

Perhaps the most obvious candidate for a belief that has significantly contributed to the environmental crisis is that the natural world has value only as a commodity, as raw material for human purposes, primarily the production of goods. This belief is associated with, but is not exclusive to capitalism, for it is common in industrial societies (and in those seeking to industrialize). As we shall see below, it is a product of the scientific revolution and the rise of a mercantile economy in the 16th and 17th centuries.

Marx objected to capitalism's treatment of human beings as mere commodities, a practice he believed was the result of private ownership of the means of production. While Marx and his followers had little to say against the exploitation of nature, Herbert Marcuse (1972) has recently extended Marx's analysis to include the natural world. *"Nature, too, awaits the revolution!"* he wrote, thus placing the natural world within the class of the oppressed and in need of liberation. Marcuse took the approach of radical ecologists and asserted that nature has intrinsic value, that all things exist "for their own sake" and not simply for us. Hence neither the natural world nor human beings can be regarded as mere commodities.

Does Marcuse's view of our relationship to nature mean that industrial society must be scrapped along with the belief that nature has value only as a commodity? How can we liberate the natural world and continue to appropriate it for our use? The issue is not simply doing away with industrial society; it is finding a way to live that does not treat nature merely as a commodity. Of course we must use the natural world to grow food and make products for shelter, transportation, recreation, etc. In that sense trees are commodities, but the crucial distinction here is to not allow them to become merely commodities.

Immanuel Kant (1785) made this distinction a fundamental feature of his ethical thought. Persons, he claimed, have intrinsic value; they are ends-in-themselves. Consequently they cannot be treated merely as a means to someone else's project. While we all serve as means by which other pursue their projects, no moral harm is done unless we are treated merely as a means; in other words, like a commodity, having value only to the extent that we serve as a means. In more modern language, commodities cannot be the bearers of rights, and to treat persons as commodities is to deny their rights.

This critique of nature as a commodity raises a fundamental ethical question: Does it make sense to grant rights to trees, animals, rivers, rocks, and ecosystems? Philosophers have addressed this question at some length; it is one of the ideas that can contribute to a better relationship to the natural world and will be examined in that section of the paper.

Patriarchy

Some of the most interesting and controversial ideas in philosophy in recent years have been found in the feminist critiques of dominant institutions and their justifications. They have made a strong case for the many pervasive and subtle ways our society and our thinking are influenced by masculine (i.e. white, Western, male) conceptions. Much of what had been taken for granted as natural and reasonable has been shown to reflect male conceptions and concerns.

It is not surprising that the women's liberation movement has provided insights into the domination of nature. While Marcuse assimilated nature to the exploited workers, feminists like Ynestra King (1983) see nature as a female victim of male domination. Just as Marx viewed capitalism as treating both workers and the natural world as commodities, many feminist thinkers link the domination of nature to the domination of women.

The key to this connection is the conception of nature as a nurturing, female presence. Before the beginnings of the modern world in the seventeenth century the image of "Mother Nature" was understood in a more literal way and tended to restrain human treatment of nature. Carolyn Merchant's study of mining details the ethical problems of digging into a living, female organism, the earth. *"Miners offered propitiation to the deities of the soil and subterranean world, and observed strict cleanliness, sexual abstinence, and fasting before violating the sacredness of the living earth by sinking a mine."* Debates about the ethics of mining were common in sixteenth century and reflected conflicts between the older, organic cosmology and the demands of the emerging mercantile system.

This attitude to nature was replaced with one which viewed nature as, at worst, wild and threatening, and at best, deceptive and secretive. In either case, force and guile were needed to extract nature's secrets rather than sacred rites. Francis Bacon's writings on science are full of such images, and feminist writers have pointed out the parallels between the mistreatment of women and the "rape" of "virgin" land (Kolodny, 1975; Gray, 1981; Marietta, 1984).

The domination of nature was the promise of the new science, and it is hardly surprising that women, long identified with nature by virtue of their procreative and nurturing function, were included.

"[Woman] became the embodiment of the biological function, the image of nature, the subjugation of which constituted that civilization's title to fame. For millennia men dreamed of acquiring absolute mastery over nature, of converting the cosmos into one immense hunting ground.

It was to this that the idea of man was geared in a male-dominated society.
This was the significance of reason, his proudest boast".

(Horkheimer and Adorno, 1972, p. 248).

Nancy Chodorow, (1978) a feminist psychiatrist, traces this idea to the different ways boys and girls achieve differentiation from their mothers. For boys, separation and differentiation from the mother is essential for gender identity. In contrast, girls develop their gender identity by being like their mother. As a result, Chodorow concludes, males begin with a social orientation that is positional, while for females it is personal. Men thus tend to see success in competitive and individual achievement (where control over the environment is essential), while women are oriented toward relationships.

The work of Gilligan (1982) revealed that even the moral language used to assert the claims of nature reflects a masculine bias. Feminists point out that the idea of "rights," has its origins in situations that emphasize individuality, competitiveness, and achievement. A right presupposes a context of conflict and competition in which rights protect participants. Further, as Cheney (1987) has argued, these ideas are tied to notions of individuality and achievement by overcoming obstacles, all of which contribute to an exploitative attitude toward nature.

Women, according to Gilligan, tend to see moral issues not in terms of rights and obligations but of preserving and nurturing relationships and connections. Women learn to see themselves in relationships with responsibilities for care, whereas men establish themselves in terms of individual achievement. A striking example of this was found in the study of how boys and girls play. Boy's play is typically interrupted by frequent squabbles about rules and infractions (rights and obligations). In contrast, girls tended to stop play whenever conflict arose, in order to preserve relationships which might be harmed.

The result is that many feminists see concern for nature weakened by the introduction of a moral language which is drawn from individuality, competition and hierarchy - values that feed the exploitation of nature. In place of the rights of nature, feminists propose a more holistic sense of our place in the great web of life. Elizabeth Dodson Gray (1981) describes this alternative way of valuing nature: *"The new understanding of life must be systemic and interconnected....the reality of life on earth is a whole...in which everything has its part to play and can be respected and accorded dignity. "*

The message from ecofeminists is that we can learn much about the exploitation of nature by looking at parallels with the exploitation of women. An essential contributor to both is a set of masculine values that emphasize achievement through individual competition sanctioned by rights. They recommend that we rethink our relationship to the natural world, first by abandoning the masculine language of ethics with its underlying conceptions of individuality, rights, and hierarchy, and second by adopting a more ecological mode of thought which gives prominence to connectedness and the need to accept, nurture, and preserve other life forms.

In a world where exploitation is commonplace, it is not clear that the language of rights should be abandoned. One of the primary functions of rights is to protect, to set a limit to what can be done by others. It is clearly better to have relationships based on trust, acceptance and caring, and where conditions allow them to flourish we can - and should - abandon the concepts of rights and obligations. But where those conditions do not obtain, it would be folly to not use the powerful language of the French Revolution and the American Declaration of Independence. This is our most promising path to bring the natural world into the moral community, to extend our familiar moral concepts to new subjects. Historically, this is how our conception of the moral community has grown, by extension. The parallel with black slaves in the USA is particularly important and will be discussed later.

Christianity

Many thinkers believe that Christianity - or a particular version of it - contributed to the environmental crisis. By drawing a strong distinction between humanity and nature, with the idea that nature was given by God solely for our benefit with the right of dominion, Christianity denied the intrinsic value of the natural world. In the account of creation in Genesis, God gives dominion over the earth to humans, and while those texts need not be understood to mean that the world is our property (i.e., a commodity), that view has dominated popular thinking. Like the change in the conception of nature that occurred in the 16th and 17th centuries, theologies that were more in harmony with expanding mercantilism also appeared at this time.

Equally important, the world given by God was something entirely separate; its rigid monotheism rejected any sort of animistic spiritualism in nature. No divine presence was found in nature to animate it and restrain human use of it. God was other than the world and left it to our management.

Finally, as White (1967) and Berman (1981) point out, the status of this world in this version of Christianity is decidedly second-class. It is not our true home, which is in heaven. Earth is only a temporary stop on our way to eternity, a "halfway house of trial and testing." It would be a mistake to focus attention on preserving the natural world because our sights should be upon Heaven.

These are not the only ways the Christian message can be read, and Nash (1989) devotes a chapter to ecotheologians who have fashioned much more environmentally friendly theologies than the one sketched here. But these Christian ideas have significantly contributed to the mistreatment of the earth and must be replaced if we are to save what is left.

II

For whom are we saving the earth?: for us, for our children, and for the earth itself. Each of these answers raises interesting philosophical issues.

The Appeal to Self Interest

This is the idea of choice for the majority of writers, probably because it is seen as the most widely shared. Whether we are Christians or Moslems, Marxists or capitalists, we are all interested in what happens to us. The message of these writers is: environmental problems will adversely affect your life, therefore you should act to solve them. Citizens and policy makers are urged to take action on the grounds that the quality of their lives is now or soon will be degraded. The hope is that people hearing this message will subjectively appropriate the truth that their health, for example, is at risk and be willing to both make changes in their personal behaviour and support political efforts to deal with the problem. This approach includes the older conservationist philosophies which advocated wise management of "natural resources." They advocate concern for the natural world out of self-interest and economic efficiency.

One danger of the appeal to self-interest is that it encourages sensationalistic methods to drive home the danger. In order to get people to see the problem, many believe it must be presented as dramatically as possible. These appeals have a messianic fervour as the authors grab us by our lapels and shout their cases. But this global warming of public discourse only dulls sensitivity. Like advertising, messages of impending doom are a constant feature of contemporary life and the only sane policy is to generally tune them out.

A more substantial reservation about appeals to self interest is that it isn't obviously true that particular lives are going to be affected. Less ozone means that there is a higher risk of skin cancer; organic chemicals in the drinking water increase the risk of intestinal diseases; and acid rain will spoil vacation spots. All true, but not necessarily true for particular persons. I may not develop any of these disorders or care about Bill McKibben's Adirondack forests. Or I may add a bit of sun screen, drink bottled water, and take winter vacations in Denmark. If the focus is on the quality of my life, an appropriate response may be to ignore the danger, take the additional risk, or make changes that will lessen the impact of the problem on me.

Finding an operating public telephone in downtown Manhattan is difficult. But this has not led to action that will make operating pay phones available; instead, car phones have proliferated, allowing the affluent to solve the problem. Framing the issue as a threat to individual self-interest invites individual solutions rather than concerted efforts. Acknowledging that there is a environmental threat to my life does not necessarily lead to action to solve the environmental problem.

The Claims of Future Generations

A more attractive value that is sometimes invoked in discussions of environmental problems is the claims of future generations. This approach holds that we owe our children, and their children, an environment that is, at least, not degraded and life-threatening. It is wrong to ask future generations to pay the price for our polluting way of life with a poisoned earth and severe climatic changes. Parents traditionally take pride in passing on what is valuable and important to their children, whether it is a way of life, the family estate, religious practices, or an education. Our children will inherit our world, and just as we want to pass on to them the best of our values and thinking, so also we want to pass on to them a world that is clean attractive, and intact. We don't simply inherit the earth; we borrow it from our children.

This approach is more attractive because it appeals to our love for something other than ourselves. Another advantage is that it is not so easy to privatize the problem. Affluent families can more easily protect their children from environmental dangers, but the time frame acts against this. I may not live long enough to be touched by approaching environmental problems, but my children are very likely to be affected, and their children - my grandchildren - certainly will be. This is particularly true for issues like global warming and the poisoning of the air and water. From these dangers no one is immune; all will be affected, even the rich.

There are important philosophical questions here that a thoughtful person will raise. What, exactly, do we owe to our children, individually and collectively? Can one have obligations to nonexistent persons, like grandchildren not yet conceived? We need to have coherent and convincing answers to these questions if this appeal is to be successful.

Joel Feinberg (1980) argues that future generations have a right to a liveable environment because their interests can be affected at the present time. We know with reasonable certainty that there will be descendants of human beings who will need - and therefore have interests in - clean air, water, and natural resources. Since we can harm their interests now, by harming the environment, they have an ethical claim on us. *"The identity of the owners of these interests is now necessarily obscure, but the fact of their interest-ownership is crystal clear, and that is all that is necessary to certify the coherence of present talk about their rights."* For Feinberg, the relevant ethical concept here is <u>interest</u>, not actual existence. Although future generations do not yet exist, we know that they will exist, and the interests they will have are causally connected to present actions and policies.

Feinberg uses a similar argument to explain why we feel obligated to carry out the wishes of dead persons. When a person dies, the interests of that person may continue - as when money is left for the education of children or to the work of a particular organization. Thus we acknowledge both legal and ethical obligations to persons who have ceased to exist, and we should accept our obligations to our not-yet-existing descendants for the same reason.

Feinberg sees his work as a contribution toward meeting our obligations to our children and their children:

> *"For several centuries now human beings have run roughshod over the lands of our planet, just as if the animals who do live there and the generations of humans who will live there had no claims on them whatsoever. Philosophers have not helped matters by arguing that animals and future generations are not the kinds of beings who have rights now...I have tried in this essay to dispel the conceptual confusions that make such conclusions possible. To acknowledge their rights is the very least we can do...But that is something."*

Respect for Nature

One of the most profound developments in environmental philosophy concerns the extension of traditional ethical concepts, such as rights, obligations, and responsibility, to nonhuman animals and to inanimate objects such as rocks, trees, rivers, entire ecosystems, and even the earth itself. This development has opened a serious rift among environmental ethicists, dividing them into two groups.

One group, of which Feinberg is a member, accepts the extension of ethical concepts only to sentient beings. Their argument is that only things that are aware of something bad happening to them have an interest that demands our respect. In this respect they follow Bentham, who said *"The question is not Can they Reason? nor Can they talk?, but Can they suffer?"* For this group, since trees cannot suffer or even be aware of what we do to them, they have no "interests" and hence no ethical standing. As Peter Singer (1981) explained, *"There is nothing we can do that matters to them."* According to Joel Feinberg, (1980) it is *"absurd to say that rocks can have rights...because rocks belong to a category of entities of whom rights cannot be meaningfully predicated."*

Although this group counts many animal rights activists among its members, the boundary of sentient life is also the boundary of ethical obligation. For them, we may wish to spare the Amazon rain forest for reasons of self-interest (to prevent damage to the atmosphere), or out of a sense of obligation to future generations (prevent the bad consequences of atmospheric damage from falling on our children). But we cannot, for this group, claim that the forest itself has rights that demand our respect.

Peter Singer coined the word "speciesism" to characterize persons who believed that only humans (our species) can have rights. In an ironic turn, Singer and others like him have been called "sentientists" for their refusal to extend the language of rights to non-sentient beings. A good summary of the thinking of those who reject Singer and Feinberg is presented by Willard Entemen (1990) *"If neither standard ethics nor standard economics are conceptually empowered to deal with the environment and yet if environmental issues are important, we should begin to speculate on the conceptual changes which are necessary to deal with those issues."*

The second group is characterized by its willingness to move beyond the boundaries of sentience and grant rights to features of the natural world. This is a much more diverse group which draws upon different philosophical sources. Probably the most influential takes its inspiration from Aldo Leopold's A Sand County Almanac. The philosophy inspired by Leopold and most fully developed by J. Baird Callicott (1989) is called "The Land Ethic." On this view, nonhuman life and nonliving matter have moral standing because of two considerations. The first is our dependence upon the ecological system in which we live. In Aldo Leopold's words, *"All ethics rest upon a single premise: that the individual is a member of a community of interdependent parts."* Individualistic ethical thinking fails to acknowledge the primacy of the community and focuses on the independent, autonomous individual. The holism of the Land Ethic is based on the scientific understanding of nature as a cooperative, ecological community which contains little rugged individualism. For Callicott, (1980) right and wrong primary applies to the biotic community, not to individuals: *"Oceans and lakes, mountains, forests, and wetlands are assigned a greater value than individual animals,"* including human beings. It follows that we have obligations to ecosystems and species based on the survival of the ecosystem and the continuation of the biotic community. This approach is ecocentric rather than individualistic.

The second reason which gives moral standing to nonhuman living beings and to nonliving matter is the relationship between ethical rights and intrinsic value. Respect for nature proponents invoke the principle that whatever has intrinsic worth is entitled to our respect and, hence, has rights. Just as Feinberg anchored his defense of animal rights on the basis of interests, intrinsic worth is the touchstone for the extension of rights to nature. Humans, animals, and many features of the natural world can be said to have worth apart from human purposes and interests and therefore there are limits to what others may do to harm that worth.

Celebration of the natural world for its own sake has deep roots in American naturalist thought. John Muir (1901) asked, *"What good are rattlesnakes for?"* and answered that they were *"good for themselves, and we need not begrudge them their share of life."* Muir held that rattlesnakes and redwood trees were part of the moral community, entitled to their share of life, free from our interference.

One way to acknowledge the intrinsic worth of nature is through the tradition of liberalism, which has at its core the idea that freedom for all is best protected by establishing individual rights which set limits to individual, corporate, and government actions. Liberalism, particularly in the United States, has a long tradition of protecting unpopular and/or powerless minorities from exploitation by the more powerful elements of society. Liberal ecologists seek to bring the natural world into the community.

A useful parallel is the struggle to abolish slavery. William Lloyd Garrison, the famous American abolitionist, sought to include blacks within the orbit of citizens who possessed basic rights. As fellow citizens, blacks could no longer be regarded as property; their rights would limit what others could do to them. Marcuse, and other radical ecologists, are simply proposing that nature be regarded as part of the community with appropriate rights.

Proponents of slavery argued that the economy of the South could not be sustained without slave labour, that crushing economic consequences would follow if slaves were freed, that is, given the rights of citizens. The reply of the abolitionists was simple and uncompromising: we must find new ways to live that do not violate the rights of black persons. Today many thinkers are saying this about the northern spotted owl and other parts of the natural world which are threatened by us.

CONCLUSION

The environmental crisis will require significant social and political changes. Those changes can and should be framed in terms of the values of self-interest, but more emphasis should be placed on the needs of future generations and respect for nature.

Few people are immune to what happens to our children or to the attraction of the natural world. Michael Ignatieff (1984, p. 123) and others have called attention to the impact of actually seeing the earth and experiencing its beauty and fragility against the black backdrop of deep space. *"No generation has ever understood the common nature of our fate more deeply, and out of that understanding may be born a real identification, not with this country or that, but with the earth itself."* Using some of the philosophical ideas developed above, those values can be encouraged and given a central role in the process of coming to terms with the environmental crisis.

Concern for our children and our earth are worthy values to steer us through a difficult time of adjustment to new environmental realities. Will these values move ordinary people and policy makers to make the adjustments that will be needed? I like the answer given by another environmentalist:

> *"Can we rely on it that a "turning around will b accomplished by enough people quickly enough to save the modern world"? This question is often asked, but whatever answer is given to it will mislead. The answer "yes" would lead to complacency; the answer "no" to despair. It is desirable to leave these perplexities behind us and get down to work."*

> *E.F.Schumacher (1977).*

REFERENCES

Berman, M. (1981), The Reenchantment of the World. Ithaca, NY: Cornell University Press.

Callaghan, D. (1989), What Kind of Life?: The Limits of Medical Progress. New York: Simon and Schuster.

Callicott, J.B. (1980), Animal Liberation: A Triangular Affair. Environmental Ethics 2, (Winter), pp. 311-328.

Callicott, J. B. (1989), In Defense of the Land Ethic. Albany, NY: SUNY Press.

Cheney, J. (1987), Ecofeminism and Deep Ecology. Environmental Ethics 9, (Summer), pp. 115-145.

Chodorow, N. (1978), The Reproduction of Mothering. Berkeley, CA: University of California Press.

Enteman, W.F. (1990), in Business Ethics Report: Business, Ethics, and the Environment, p. 28. Center for Business Ethics, Bentley College, Waltham, MA.

Feinberg, J. (1980), Rights, Justice, and the Bounds of Liberty, p.181, 183, 209. Princeton, NJ: Princeton University Press.

Gilligan, C. (1982), In a Different Voice: Psychological Theory and Women's Development. Cambridge, MA: Harvard University Press.

Golub, R, and Brus, E. (1990), The Almanac of Science and Technology, Chapter 7. Boston: Harcourt Brace Jovanovich.

Gray, E.D. (1981), Green Paradise Lost. Wellesley, MA: Roundtable Press.

Horkheimer, M, and Adorno, T.W. (1972), Dialectic of Enlightenment, p. 248. New York: Seabury Press.

Ignatieff, Michael (1985), The Needs of Strangers, p. 139. New York: Viking Penguin, Inc.

Kant, I. (1785), Foundations of the Metaphysics of Morals. New York: McMillan, 1985 edition.

Kierkegaard, S. (1846), Concluding Unscientific Postscript, pp. 169-182. Princeton, NJ: Princeton University Press, 1941 edition.

King, Y. (1983), Toward an Ecological Feminism and a Feminist Ecology, pp. 118-129. In Rothschild (1983).

Kolodny, A. (1975), The Lay of the Land. Chapel Hill, NC: University of North Carolina Press.

Leopold, A. (1949), A Sand County Almanac, pp. 209-210, 203. New York: Oxford University Press.

Marcuse, H. (1972), Counterrevolution and Revolt. Boston: Beacon Press.

Marietta, Jr., D. (1984), Environmentalism, Feminism, and the Future of American Society, The Humanist 44 (May-June), 1984.

McKibben, B. (1989), The End of Nature, p. 6. New York: Random House.

Merchant, C. Mining the Earth's Womb, p. 100. In Rothschild (1983). Muir, J. (1901). Our National Parks, pp. 57-58. Boston: Houghton Mifflin.

Nash, R. (1989), The Rights of Nature, Madison, WI: University of Wisconsin Press.

Rothschild, J. (1983), Machina Ex Dea: Feminist Perspectives on Technology. New York: Pergamon Press.

E.F.Schumacher, (1977), A Guide for the Perplexed. New York: Harper and Row.

Singer, P. (1981), The Expanding Circle: Ethics and Sociobiology, p. 121. New York: Farrar, Straus, and Giroux.

White, Jr., L. (1967), The Historical Roots of Our Ecological Crisis. Science 155 (March 10), pp. 1203-1207.

15.

PLURAL RATIONALITIES: THE RUDIMENTS OF A PRACTICAL SCIENCE OF THE INCHOATE[1]

MICHAEL THOMPSON
The Musgrave Institute
52 Northolme Road, London N5 2UX,
England

ABSTRACT

1. *Environmental debates reflect the existence of plural rationalities; sets of convictions about the nature of the world we live in that are fundamentally contradictory and that generate different definitions both of the environmental problems we face and of the solutions that are available to us.*

2. *These rationalities are linked to particular forms of social organisation - markets, hierarchies, egalitarian groups and excluded margins - all of which, in varying strengths and patterns of alliance, are the inescapable features of any society.*

3. *To be effective, environmental decision making must take constructive account of these rationalities.*

---

[1]   This paper summarises a line of argument that is set out in two recently published books: Schwarz and Thompson (1990) and Thompson, Ellis and Wildavsky (1990). The development of this argument in terms of varieties of uncertainty, was supported by the British Economic and Social Research Council (award reference number w 100311002).

Examples of this approach include: liquefied gas terminal siting, Himalayan deforestation, hazardous waste management, global energy futures, oil and gas reserves estimation, the health effects of low level radiation, and mass housing.

Key Words: Clumsy Institutions, Cultural Theory, Myths of Nature, Plural Rationalities, Uncertainty.

INTRODUCTION

Many important areas of human endeavour are characterised not just by uncertainty (that is, the absence of certainty) but by <u>contradictory certainties:</u> severely divergent and mutually irreconcilable sets of convictions as to how the world is and people are. I will begin by giving a number of examples, restricting myself, first, to examples that have obvious policy implications and, second, to case studies that I happen to have worked on.

1. Liquefied Natural Gas (LNG)

The technology for liquefying and transporting natural gas is recognised as potentially extremely hazardous, and much effort has been devoted to determining the scale of these hazards. Such information, for instance, is seen as crucial to the safe siting of the terminals where the ships unload their cargoes into land-based storage tanks. The key question is: how far, under favourable climatic conditions, will the vapour cloud from a large spill remain flammable? "One and a half kilometres," says one expert; "One hundred and twenty," says another (Mandl and Lathrop, 1983). If the first expert is right, the terminals can be sited almost anywhere; if the second expert is right, they can be sited almost nowhere.

Though this question is central to the viability of this technology, no test (let alone experiment) has yet been devised that will indicate which expert is right (Thompson, 1983).

2. The Himalayan Environment

The extensive deforestation that is seen as the root cause of the environmental degradation of the entire Himalayan region is conceptualized as a vicious circle; the forest is being used faster than it grows. There have been many attempts to measure these two rates, and it is an easy matter to tabulate and compare them. This exercise reveals that estimates of the per capita fuelwood consumption rate vary by a factor of 67 and that estimates of the sustainable yield from forest production vary by a factor of at least 150. Far from giving us a precise, quantitative description of the vicious cycle of degradation, these results tell us that (if the most pessimistic estimates are correct) the Himalaya will be as bald as a coot overnight and that (if the most optimistic estimates are correct) they will shortly sink beneath the greatest accumulation of biomass the world has ever seen (Thompson, Hatley and Warburton, 1987).

Again, though it might appear to be an easy matter to achieve an accurate measurement of these two rates, the physical and ecological heterogeneity of the region ensures that there is no such thing as the sustainable yield from forest production. And the subtle variations in farming system, resource perception, fuel substitution, out-migration and so on, do much the same for the fuelwood consumption rate (Ives, Messerli and Thompson, 1987, Ives and Messerli, 1989).

3. Energy Futures

What will Western Europe's future energy demands be? How will they be met? Much money and expertise has been devoted to answering these questions, and the role of solar energy in all this has emerged as one of the key issues. What contribution, the policymakers ask, can solar energy make to Western Europe's future? "At the most, 5 percent," says one group of experts; "At the least, 95 percent," says another (Caputo, 1981).

Again, the gut-response - that it should not be beyond the wit of man to discover which of these estimates is right - is mistaken. The largest and most exhaustive energy modelling exercise the world has ever seen, involving 250 scientists and taking nine years to complete, has now been shown to have been "captured" by just one of the contradictory certainties it aspired to decide between (Keeepin, 1984, Thompson, 1984, Wynne, 1984).

4. Hazardous Wastes

How much hazardous waste is being produced each year in the United States? This is the first question the Environmental Protection Agency (the EPA) asks, once it realises it has a hazardous waste problem on its hands. One authoritative report says "40 million metric tons"; "250 million," says another (Dowling, 1984, Wynne, 1988). Both reports, moreover, concede that they only take the reported (that is the legal) waste into account (and that they have to assume that all the reported waste is, in fact, hazardous). Illegal disposal, they concede, does happen but they have no way of telling how much is being handled in this way (nor can they estimate the scale of the over-reporting that, they know, is also going on). So each of these distressingly divergent figures needs the suffix "plus or minus as much as you care to add or subtract."

Again, it is difficult to see how these sorts of uncertainty bounds can ever be narrowed, and the obstacles are compounded once we realise that the hazard is a property, not of the waste itself, but of it and the environment into which it enters. The obstacles are further compounded once we realize that the definition of what is and is not waste is not inherent to the material itself but is a quality that is conferred on (and withdrawn from) it by processes that are entirely social (Thompson, 1979, 1986, 1987a).

5. Oil and Gas Reserves

For more than fifty years now, the United States Department of Energy, armed with formidable legal powers to force the oil companies to reveal their data, has been trying to determine how much oil and gas there is "down there" (this information being seen as crucial for the effective regulation of the industry). Estimates have consistently varied by a factor of between 12 and 14 and have shown no tendency to converge, even as more and more of the historical answer (oil and gas that was "down there" and was brought "up here") has been revealed (Wildavsky and Tenenbaum, 1981).

These estimates, moreover, far from being spread out across the range in a normal-type distribution, are clustered into just three contradictory certainties: the optimist's, the moderate's and the conservative's, as they have been called. These three certainty clumps, which have persisted for half a century, become even more remarkable when we realize that (unlike, say, the case of Himalayan fuelwood) there is nothing that can be measured (Schwarz and Thompson, 1990). The whole estimation process, as its historian, Schanz (1978), has observed, is equivalent to *"going to an unfamiliar supermarket on a foggy night and trying to estimate the total amount of asphalt used in paving the parking lot, with no other data than a cubic inch sample of the blacktop used."*

6. The Health Effects of Low Level Radiation

What is the shape of the dose/response curve at low levels of ionizing radiation? At high levels, of course, the answer is well known: the relationship is linear (we know this from animal experiments, from human accidents and from follow-up studies to the Hiroshima and Nagasaki atom bombs) but does this relationship also hold for low levels? "Yes," say the scientists from the British Department of the Environment, and the safe doses and other standards are set accordingly. But other experts (the US organisation, Scientists and Engineers for Secure Energy, for instance) argue that the curve is quadratic, defining a threshold below which no harm is inflicted. Still other experts (Alice Stewart of Birmingham University, for instance) argue in the opposite direction. The curve, they insist, is parabolic, causing proportionately greater harm at low levels of exposure than at higher ones (Schwarz and Thompson, 1990).

If the quadraticists are right then (provided nuclear technology is engineered below the threshold) there will be nothing to worry about. If the parabolists are right then the more nuclear technology we have the more harm will befall us. If the linearists are right then the technology will be "neutral": since the risks it brings are not inherently one way or the other, they can (and should) be planned, managed and regulated. The acceptability of the entire technology, therefore, hinges on which of these contradictory certainties is the right one. Unfortunately, to find that out you would have to try to give tumours to more mice than there are atoms in the universe! And then there would still be the vexed question of the extrapolation from rodents to humans!

SO WHAT?

The first thing that must be said about this "contradictory certainties" approach is that it is not (as some of those on its receiving end have claimed) an attempt to rubbish science. In pointing out that every penny that is spent in trying to find out something that is demonstrably unknowable (how much oil and gas is "down there", for instance) or incapable of existing (the fuelwood consumption rate, for instance) is a penny wasted, this approach is not attacking science; it is attacking stupidity (and sometimes, dishonesty) dressed up to look like science. How, we must ask, have so many distinguished scientists and so many august institutions made such fools of themselves?

The answer, at bottom, is that they have strayed out of the choate (the realm in which facts "speak for themselves") and into the inchoate (the realm in which things are so ill-formed that they simply cannot do what a fact has to do). Science, by shrewdly restricting itself, first, to questions that are worth asking and, second, to questions to which it is likely to be able to get an answer, has systematically expanded the realm of the choate at the expense of the inchoate. That is science's great and continuing achievement. But, in each of the examples I have given above, the scientists have not expanded the realm of the choate; they have leapt right out of it! In Alvin Weinberg's terminology, they have unwittingly moved out of science and into trans-science (Weinberg, 1972).

My argument is not that Weinberg 's trans-science is impossible - something to be avoided like the plague - but that it is very different from the sort of science (perhaps we should call it cis-science: this side of the choate/inchoate boundary), we have long been used to. It is the science of the in-choate, and its aim is not to determine which contradictory certainty is the right one but to explain the emergence and persistence of the contradictory certainties themselves. Where the science we are used to progresses by asking "What are the facts?", the science of the inchoate asks "What would you like the facts to be?".

Scientists and policymakers, however, recoil from asking this second question because it seems to open the floodgate of unconstrained relativism. If they grant credibility to more than one definition of what the problem at issue is, and if they concede that people make the world to fit the lives they wish to lead (rather than the other way round), there will, they fear, be no end to it all. Once they take that fateful first step from one reality to two, the brakes will be off and infinity will be the next stop. An over-restrictive insistence on just one state of the world, they feel, is far preferable to what they see to be the only alternative: As many plausible states as there are people to make them plausible. This, of course, is the realist/relativist impasse that so bedevils the philosophy of knowledge. It is, however, a phoney impasse. The relativism, though it is certainly there, is so severely constrained that, far from shooting off to infinity, it quickly settles down around the number four.

WHY FOUR?

The most elegant route to the four constructions of reality is by way of natural resource ecology. Ecologists who study managed ecosystems, such a forests, fisheries and grasslands, encounter the managing institutions not as organised arrangements of human beings but as sets of interventions in those ecosystems. Time and again they have found that different managing institutions, faced with exactly the same kind of situation, do remarkably different things (some, for instance, start spraying the forest with insecticide; others stop). But they do not do just anything; there is definitely some consistency to their behaviour. The problem the ecologists faced was this: If the institutions were irrational there would be no consistency to their behaviour; if they were rational they would all end up doing the same.

The ecologists' solution (Holling, 1979, 1986, Timmerman, 1986) was to ask themselves a simple question: What are the minimal representations of reality that would have to be ascribed to each managing institution for it to be granted the dignity of rationality? Four distinct interpretations of ecosystem stability were needed, each of which could be represented by a little picture of a ball in a landscape (Figure 1).

They called these minimal representations <u>myths of nature,</u> defining a myth as a cultural device that captures, in simple and elegant form, some essence of experience and wisdom. The myths of nature, therefore, are not falsehoods; they are partial truths. And each of them, as we will see in a moment, is partial to a particular <u>way of life</u>: A particular pattern of social relationships and a particular set of moral justifications for the superiority of that pattern over other, rival, patterns.

<u>Nature Benign</u> gives us global equilibrium. Such a world is wonderfully for

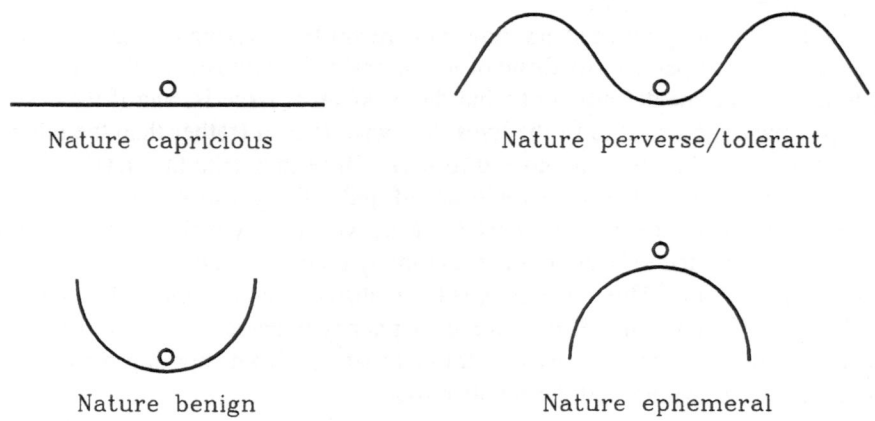

Figure 1. The myths of nature.

giving: No matter what knocks we deliver the ball will always return to the bottom of the basin. The managing institutions can therefore have a laissez-faire attitude. Nature Ephemeral is almost the exact opposite. The world, it tells us, is a terribly unforgiving place and the least jolt may cause its catastrophic collapse. The managing institutions (and everyone else) must "tread lightly on earth". Nature Perverse/Tolerant, though it may look like a cross between the first two, is quite different. Its world is one that is forgiving of most events but is vulnerable to an occasional knocking of the ball over the rim. The managing institutions, therefore, must regulate against unusual occurrences. And, if they are to distinguish between the usual and unusual, they will have to ensure that they have highly skilled and knowledgeable experts. Without them they will not be able to secure "sustainable development". Lastly, Nature Capricious is a random world. Institutions with this view of the world do not really manage, nor do they learn. They just cope with erratic events.

Each of these views of nature, and the actions it justifies, appears irrational from the perspective of any other but, nevertheless, each actor is perfectly rational, given his or her convictions as to how the world is (and given science's current inability to determine the facts that would decide the matter). The situation is one of plural (but very far from infinite) rationality, and the question it prompts us to ask is "How is each actor given his or her convictions?"

FOUR MYTHS, FOUR WAYS OF LIFE

The four myths of nature identified by ecologists map onto the typology of social relationships that has been developed by the anthropologist Mary Douglas and her co-workers (Douglas, 1978, 1982; Thompson, 1983; Gross and Rayner, 1985; Thompson, Ellis and Wildavsky, 1990). This typology is based on the answers to two central and eternal questions of human existence: "Who am I?" and "How should I behave?". Personal identity, it is argued, is determined by individuals' relationships to groups. Those who belong to a strong group - a collective that makes decisions binding on all members - will see themselves very differently to those who have weak ties with others and therefore make choices that bind only themselves. Behaviour is shaped by the strength of social prescriptions (the grid dimension) that an individual is subject to: a spectrum which runs from the free spirit to the tightly constrained. These two "dimensions of sociality", as they are called, generate four basic, and stabilisable, forms of social relationship. And, in each instance, just one of the plural rationalities can do the stabilising (Figure 2).

Two of these "archetypes" - individualists and hierarchists - are already familiar to social scientists. Indeed, the sociologist, Max Weber (1958), the political scientist, Charles Lindblom (1977) and the institutional economist, Oliver Williamson (1975), are only three of the scholars who have based entire bodies of theory on this distinction between markets and hierarchies, and the accompanying observation that each promotes a distinctive form of rationality that legitimises and enables its operation.

Market cultures stress the autonomy of individuals and their resulting freedom to bid and bargain with each other: They have a substantive rationality. The "bottom line" is what they care for, not the relational niceties of the people who happen to have come together to achieve that result. Hierarchies are made up of bounded social groups, each of which is in an orderly and ranked relationship with each other. Their attempts to coordinate these components, without violating status differentials, create a procedural rationality that is more concerned with the proprieties of who does what than with trying to evaluate the outcome (if there is one).

But this is an inadequate taxonomy. Many people reject both the individualism of the market and the inequalities of the hierarchy: They prefer the equalitarian groups of my diagram. They have a communal and critical rationality, which stresses the importance of fraternal and sororal co-operation, and therefore strives for social relationships that are voluntaristic and egalitarian. But, since this desired state of affairs

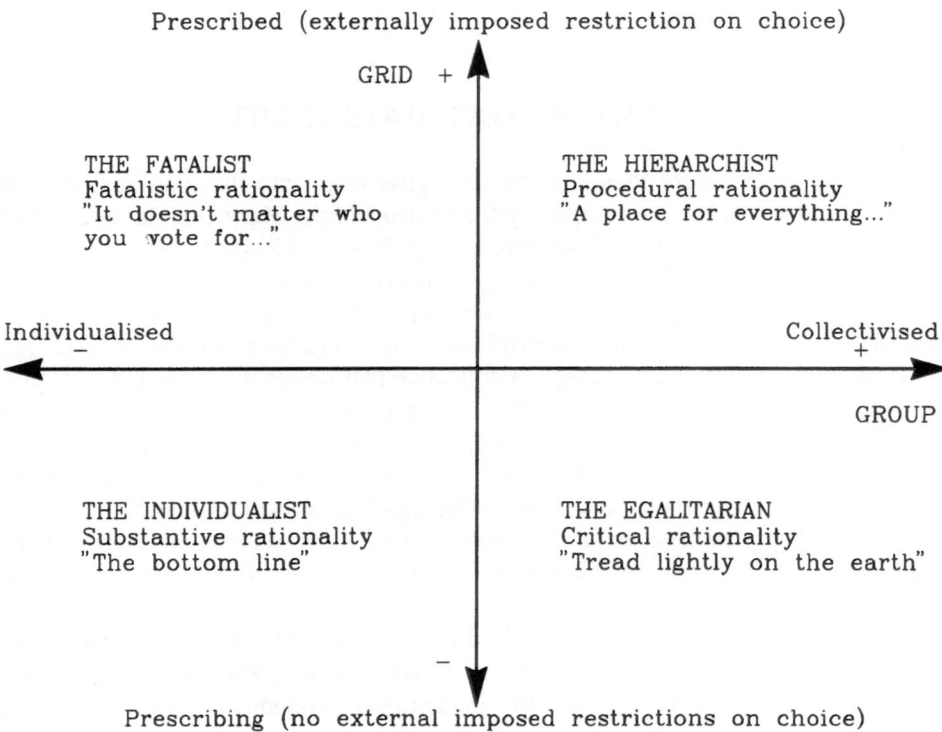

Figure 2. The two dimensions of sociality and the four rationalities.

is always threatened by the encroachment of hierarchy (which brings status differences) or by excessive individualism (which all too easily introduces inequalities of wealth, power and knowledge), collective identity has all the time to be sustained by a shared and strident criticism of what goes on outside the group. Historically, this rationality has been the driving force of socialism (but, as that movement has grown, it has been increasingly diluted by hierarchy and political entrepreneurialism) and today it is alive and well as the preferred organisational form of the Greens (and of many single-issue, public interest groups in the USA).

It is also a cruel travesty to describe all those who are individualised as bustling and untrammelled entrepreneurs: As paid-up members of "the enterprise culture". Many have so many prescriptions on their behaviour that they have minimal freedom or choice: For example, the unemployed, trailing from one welfare centre to another ad infinitum. These are the marginal members of society - the fatalists - whose inability to influence events this way or that engenders a fatalistic rationality in which outcomes, good or bad, are simply to be enjoyed or endured, but never achieved.

Each of these rationalities, when acted upon, both sustains and justifies the particular organisational form that goes along with it. The high-rise system-built tower block, for instance, is the hierarchist's solution to the housing problem; gentrification, the individualist's; cooperative self-build, the egalitarian's; homelessness, the fatalist's (Thompson, 1987b). Hierarchists trim and prune social transactions until they fit neatly into their orderly ambit, individualists pull them into the marketplace, egalitarians strive to capture them into a kind of voluntary minimalism (which, to those on the outside, often looks more like "coercive utopianism"), and fatalists endure with more or less dignity whatever comes their way.

We can now see how each of the myths of nature (the ecologist's explanation for "managerial heterogeneity") legitimates and reproduces certain kinds of institutional relationships (the anthropologist's cultural categories) (Figure 3).

This diagram, of course, is just the two earlier ones - the ecologist's and the anthropologist's - combined, and it is this "new synthesis" that has made the science of the inchoate possible.

The New Synthesis

The world of Nature Benign is most hospitable to individualists. As long as we all do our individualistic, exuberant things, a "hidden hand" (the uniformly downward slope of the landscape) will lead us to the best possible outcome. Since restrictions on individual freedom, and therefore on experimentation, would impede the attainment of this outcome, the myth of a benign nature furnishes a powerful moral justification for these particular modes of acting and learning. If we take, for example, the topical issue of hazardous waste management, Nature Benign would indicate that a sharpening of market incentives (transferable "rights to pollute", brokers to reduce the transaction and information costs of connecting some firms' waste-streams into other firms' feedstocks, self-policing to increase consumer confidence, etc.) is the way to go.

By contrast, an ephemeral nature suits egalitarian groups very well. Their small-scale organisations tread lightly upon our fragile earth, and they are only too

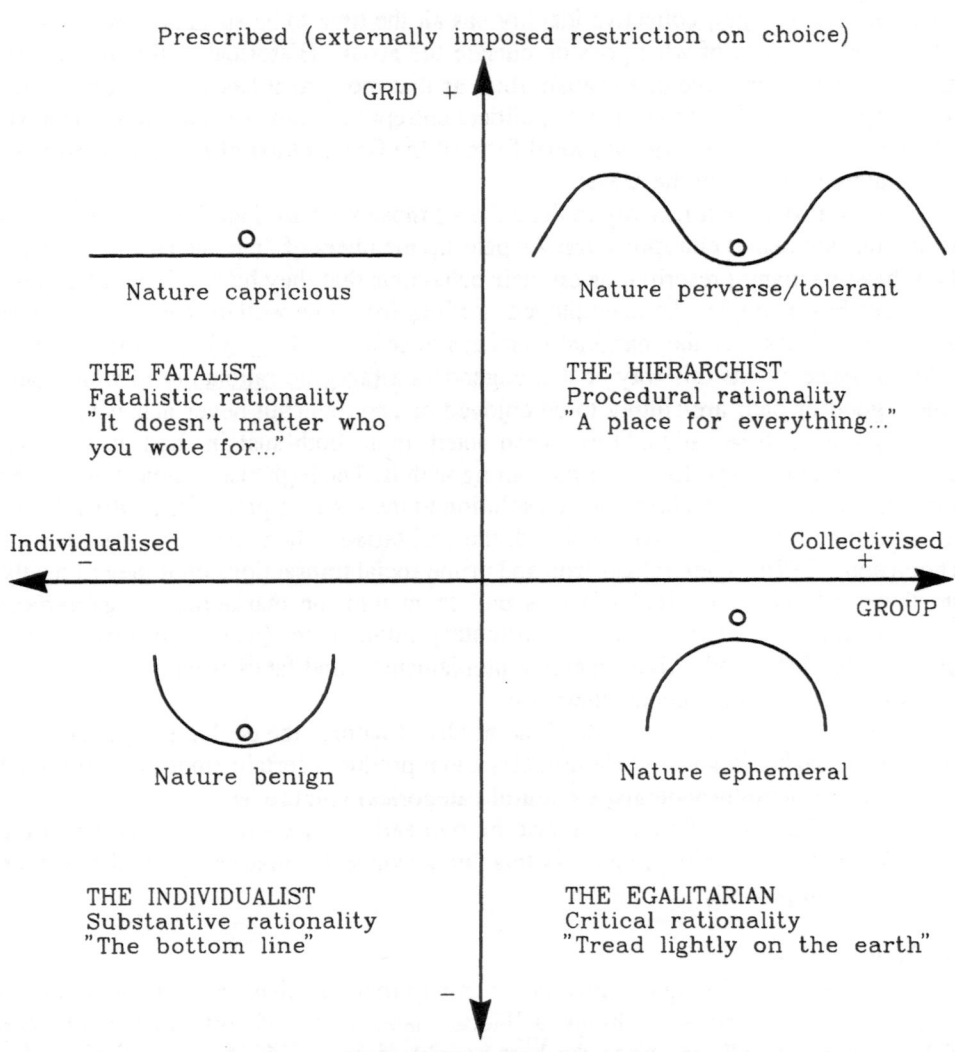

Prescribed (externally imposed restriction on choice)

GRID +

Nature capricious

THE FATALIST
Fatalistic rationality
"It doesn't matter who
you wote for..."

Nature perverse/tolerant

THE HIERARCHIST
Procedural rationality
"A place for everything..."

Individualised
−

Collectivised
+

GROUP

Nature benign

THE INDIVIDUALIST
Substantive rationality
"The bottom line"

Nature ephemeral

THE EGALITARIAN
Critical rationality
"Tread lightly on the earth"

−

Prescribing (no external imposed restrictions on choice)

Figure 3. The myths of nature mapped onto the rationalities.

happy to reeducate those who, in persisting in stamping around wildly, threaten the destruction of the entire planet. Minimal perturbation becomes the overriding moral imperative, and small becomes beautiful. Trials can go ahead only if it is certain there will be no errors. By these criteria many of the products of our consumer society are not just unnecessary, they are actually destroying the one earth that should be our most sacred trust. The myth of Nature Ephemeral tells us that there will have to be radical change now, before it is too late. Since most hazardous wastes are discharged into the environment from the production systems that, directly or indirectly, give us all these products that we do not need and should not have, the solution is an outright ban (or, better still, a consumer rejection) on all unnecessary products: A solution that has the added advantage of bringing us much nearer to the desired future - harmony with nature.

Nature Perverse/Tolerant requires strong controls to ensure that the ball never crosses the rim. And to apply those controls effectively you need precise knowledge of the line between equilibrium and disequilibrium. Neither the unbridled experimentation that goes with the zone of equilibrium, nor the timorous forbearance that accompanies the zone of disequilibrium can command much moral authority here. Everything, rather, hinges upon mapping and managing the boundary line that separates these zones. Complete know-ledge, certainty and predictability, generated by and for those whose preeminent task is to keep each mode of action - social sanctions and individual experiments - in its proper place, becomes the dominant moral concern. The situation cries out for hierarchy: Sober, expert and, above all, enduring. Only then can we have an orderly solution matched to the time-scales and complexities of the problem: Standard-setting, cradle-to-grave materials accounting systems, trip-tickets, site licences, spot-checks and precisely detailed lists of hazardous wastes.

Nature Capricious is the natural habitat for those with neither standing nor influence in society. In the other three rationalities, learning is possible (though each is disposed to learn different things) but in the flatland of Nature Capricious there are no gradients to teach us the difference between hills and dales, up and down, better and worse. Life is, and remains, a lottery. The world does things to you while you do nothing to it. All you can do is try to cope, as best you can, with a situation over which you have no control. Though those who find themselves attached to this myth produce no policies for the management of hazardous wastes, they are by no means irrelevant to those policies that are produced. They are the great risk-absorbers, enduring with more or less dignity, greater or lesser ignorance, whatever comes their way: A social sponge that the active policy-makers, in their different ways, publicly wring their hands over and privately make good use of. Without the passive risk-absorbers, (and the contradictory claims that are always made on their behalf), the rest of us would not be able to get any of our preferred policies to work.

It is by teasing out these rationalities that we can begin to make sense of what is going on in all those policy debates that are characterized by contradictory certainties. Though this approach is saying that knowledge is socially constructed, it is not saying that the world can be any way we want it to be. It is not saying that we can know nothing; only that we cannot know everything and that, within that uncertain and inchoate region, it is our institutions - our diverse patterns of social involvement - that lead us to grant credibility to one possible state of affairs rather than another.

FROM RIGHT THINKING TO RIGHT ACTING

Since each of these states of affairs is possible (both physically and socially) policies
that are nicely optimised around just one definition of what the problem is - around
just one of the myths of nature - are denied access to all the wisdom and experience
that are captured by the other three myths. Such policies are highly surprise-prone,
since they assume that nature is just one of four possible ways. They are also unlikely
to receive widespread consent, since they deny the validity of the problem and solution
definitions held by all those who happen not to subscribe to the chosen myth.

A good policy, by contrast, would be fudged and budged in such a way as to
give some recognition to each definition of problem and solution. Such a policy - a
clumsy institution, as it has been dubbed (Schapiro, 1988) - could then have this or
that component strengthened or downgraded as Mother Nature revealed more of her
hand, all the while enjoying a much higher level of consent than would any of the
much more gainly policies that are generated by those who insist on first reducing the
debate to a uni-rational frame.

REFERENCES

Caputo, R. (1981), The role of risk perception in establishing a rational energy policy
for W.Europe. In Risk: A Seminar Series (H. Kunreuther, ed.), pp. 483-514. International Institute for Applied Systems Analysis, A2361 Laxenburg, Austria.

Douglas, M. (1978), Cultural Bias. London, Royal Anthropological Institute
(Occasional Paper No. 35).

Douglas, M. (1982), Essays in the Sociology of Perception. London: Routledge and
Kegan Paul.

Dowling, M. (1984), The listing and classifying of hazardous wastes. Working Paper
84-26, International Institute for Applied Systems Analysis, A2361 Laxenburg,
Austria.

Gross, J. and Rayner, S. (1985), Measuring Culture. New York, Columbia University
Press.

Holling, C.S. (1979), Myths of ecological stability. In Studies in Crisis Management
(G. Smart and W. Stansbury, eds.) Montreal: Butterworth.

Holling, C.S. (1986), The resilience of terrestrial ecosystems. In Sustainable Development of the Biosphere (W. Clark and R. Munn, eds.) Cambridge: Cambridge University Press.

Ives, J.D., Messerli, B. and Thompson, M. (1987), Research strategy for the Himalayan region. Mountain Research and Development 7,3: 332-344.

Ives, J.D. and Messerli, B. (1989), The Himalayan Dilemma: Reconciling Development and Conservation. London: Routledge.

Keepin, B. (1984), A technical appraisal of the IIASA energy scenarios. Policy Sciences 17,3: 199-276.

Lindblom, C.E. (1977), Politics and Markets. New York: Basic.

Mandl, C. and Lathrop, J. (1983), LEG risk assessment: experts disagree. In Risk Analysis and Decision Processes (H.C. Kunreuther et al.). Berlin: Springer.

Schanz, J.J. Jr. (1978), Oil and gas resources: welcome to uncertainty. Resources 58. Reprinted in Wildavsky and Tenenbaum (1981), pp. 327-358.

Schapiro, M.H. (1988), Judicial selection and the design of clumsy institutions. Southern Californian Law Review 61, 6: 1555-1569.

Schwarz, M. and Thompson, M. (1990), Divided We Stand: Redefining Politics, Technology and Social Choice. London: Harvester Wheatsheaf.

Thompson, M. (1979), Rubbish Theory. Oxford: Oxford University Press.

Thompson, M. (1983), A cultural basis for comparison. In Risk Analysis and Decision Processes (H.C. Kunreuther et al.). Berlin, Springer.

Thompson, M. (1984), Among the energy tribes: a cultural framework for the analysis and design of energy policy. Policy Sciences 17,3: 321-339.

Thompson, M. (1986), Hazardous waste: what is it? Can we ever know? If we can't does it matter? In Safety Evaluation and Regulation of Chemicals III (F. Homburger, ed.). Basel: Karger, pp. 230-236.

Thompson, M. (1987a), The management of hazardous wastes and the hazards of wasteful management. In Waste Through the Ages (W. Schenkel, ed.). München: Verlag J. Jehle. Republished (1990) in Bradley, H (editor) Dirty Words: Writings on the History and Culture of Pollution. London: Earthscan.

Thompson, M. (1987b), Welche Gesellschaftsklassen sind potent genug, anderen ihre Zukunft aufzuoktroyieren? Und wie geht das vor sich? In Design der Zukunft (L. Burckhardt, ed.). Berlin: International Design Centre/Dumont.

Thompson, M., Ellis, R. and Wildavsky, A. (1990), Cultural Theory. Boulder Co: West View.

Thompson, M., Hatley, T. and Warburton, M. (1987), Uncertainty on a Himalayan Scale. London: Milton Ash, Ethnographica.

Timmerman, P. (1986), Mythology and surprise in the sustainable development of the biosphere. In Sustainable Development of the Biosphere (W. Clark and R. Munn, eds.). Cambridge: Cambridge University Press.

Weber, M. (1958), The Protestant Ethic and the Rise of Capitalism. New York: Free Press.

Wildavsky, A. and Tenenbaum, E. (1981), The Politics of Mistrust: Estimating American Oil and Gas Resources. Beverley Hill: Sage.

Williamson, O. (1975), Markets and Hierarchies. New York: Free Press.

Weinberg, A. (1972), Science and trans-science. Minerva 10: 209-222.

Wynne, B. (1984), The institutional context of science, models and policy: the IIASA energy study. Policy Sciences 17.3: 277-320.

Wynne, B. (1988), Risk Management and Hazardous Waste: Implementation and the Dialectics of Credibility. London: Springer.

16.

THE FORMATION OF GREEN PARTIES ENVIRONMENTALISM, STATE RESPONSE, AND POLITICAL ENTREPRENEURSHIP

EVERT VEDUNG
Uppsala University
Department of Government
P.O. Box 514, S-751 20 Uppsala
Sweden

ABSTRACT

Ideas and institutional response to ideas count but political entrepreneurship decides. New parties are formed not only through the emergence of new ideas, but also through a combination of new ideas and unfavorable responses to new ideas by existing state institutions. The role of political entrepreneurs is crucial. When the state and, particularly, major conventional parties fail to provide acceptable forms of linkage between the electorate and the government, a climate favourable towards for the formation of new parties is created. If, in these situations, there are risk-prone political entrepreneurs willing to shoulder the heavy burden of organization-building, new parties will be formed. Underlying the formation of Green parties in the 1970's and 1980's is the wave of environmentalism that has swept the world; and the perceived unfavorable response to this wave of new ideas by the state, particularly the traditional frozen party systems. The triggering factor, however, is the availability of risk-taking political entrepreneurs who are willing to make the necessary sacrifices in order to organize a new party. Behind the formation of the Green Party in Sweden in 1981 are at least three perceived failures on behalf of the conventional five-party system, all of which are related to their responses to environmentalism, or more particularly, antinuclear sentiments: 1) the failure, after 1976, of the admittedly antinuclear Center Party to discard nuclear energy, 2) the politicking of the Social Democrats, the Liberals and the Moderates in order to save face in the 1980 referendum, and 3) the alleged betrayal of the antinuclear cause by the Center Party and the Left-Party Communists, through their promise to abide by the result of the plebiscite. The situation created by these perceived failures was used by a political entrepreneur to form a party.

Key Words: Center-Periphery Dimension, Environmentalism, Freezing of Party Systems Theory, Growth-ecology Dimension, Left-Right Dimension, Nuclear Energy, Political Entrepreneurs, Rise of Green Parties, Swedish Environment Party, Swedish Nuclear Referendum.

INTRODUCTION

The Swedish September 1988 election brought about a political sensation. For the first time in seven decades, a new party, the Greens, managed to clear all constitutional barriers and enter Parliament. The age-old Swedish five-party pattern was broken.

1. THE EMERGENCE OF GREEN PARTIES

In a wider international context, however, the feat of the Swedish Greens was less spectacular. In the 1970's and 1980's, environmental parties were established throughout the industrialized world, including Japan. While tiny in most countries - between 3 to 10 per cent of the electorate - some of them succeeded in gaining seats in the national parliaments.

An overview over the emergence of Green parties and Green lists in developed countries is provided in Table 1.

Table 1
Green Parties and Green Lists in Europe and some other OECD-countries.

Emergence	Country	Party name
1972 (May 20)	New Zealand	Values, Green Party of Aotearoa
1973 (Feb)	United Kingdom	Green Party
1974	France	Env. movement runs presidential candidate
1974	Belgium	Green lists (Agalev and Ecolo)
1979 (Mar)	Germany (West)	Green list in the European election (SPV)
1979	Luxembourg	Alternative list (Alternative Lëscht: Wiert lech!)
1980 (Jan)	Germany (West)	Greens (Die Grünen)
1980 (Mar)	Belgium	Ecolo (Wallonie Ecologie)
1981 (Sep)	Sweden	Environmental Party (Miljöpartiet De Gröna)
1981 (Dec)	France	The Greens (Les Verts)
1981 (Dec)	Ireland	Ecology Party
1982 (Nov)	Austria	The Green Alternative (Die Grüne Alternative)
1982	Belgium	Agalev (Anders Gaan Leven)
1983 (Jun)	Luxembourg	Green Alternative (Déi Gréng Alternativ)
1983 (Aug)	Canada	Green Party
1983 (Oct)	Denmark	Greens (De Grønne)
1983 (Dec)	The Netherlands	Greens (De Groenen)
1983	Japan	Greens
1983	Finland	Green list
1983 (May)	Switzerland	Green Party (Die Grüne Partei)
1983	Switzerland	Green Alternative (Alternative Verte de Suisse)
1986 (Nov)	Italy	Federation of Green Lists (Federazione delle Liste Verdi)
1987 (Feb)	Finland	Green Union (Vihreä Liitto)

Comments. The list is compiled from Müller-Rommel (1989), Parkin (1989) and from my own correspondence. Entering only national environmental parties would be misleading. First of all, in a country like Belgium for example, there is no national environmental party because there are no national parties at all, only regional ones. Secondly, environmental movements have participated in elections without forming any parties; they have put up separate lists with candidates of their own. Thirdly, in one of the countries for a period of time, there was no national Green party but a nationwide federation of local Green parties. Therefore, in addition to Green parties proper, also alternative lists and the above federation of local Green parties are included in the Table.

Originally named Values Party, the New Zealand party adopted its current name in 1986. The British party which was originally named People, changed its name in 1975 to the Ecology Party and, later in 1985, to the Green Party. In the Belgian Parliament there are only regional parties. Agalev (short for Anders Gaan Leven) is based in Flandres; Ecolo (short for Wallonie Ecologie) is based in Wallonia. The French Greens were formed in 1984 through a merger of Les Verts - Confédération Ecologiste (founded in Dec. of 1981) and Les Verts - Parti Ecologiste (founded in Nov. of 1982). The Irish party substituted the name Ecology Party in 1983/84. The Austrian Green Alternative, established in September of 1986, can be traced back to the Austrian Alternative List and the United Greens of Austria, which was founded in November and December 1982, respectively. The Green Party of Switzerland was founded in May of 1986 but dates back to the Federation of Green Parties (Föderation der Grünen Parteien) formed in 1983.

The pioneering parties of an environmental persuasion were created in the beginning of the 1970's. The world's oldest environmental party, the Values Party, was founded in New Zealand in 1972. A year later, the first West European Green party, People, was formed in Britain. One year after that, the French environmental movement ran a candidate, the famous agronomist Professor René Dumont, in the presidential election. He is regarded as the first Green candidate in Europe in a general balloting.

In 1979, the environmental movement in West Germany created a list of its own in the European election and, in January of 1980, the Green Party was founded. In 1983, the party received 5.6 per cent of the vote, thereby clearing the five per cent hurdle, and gained 27 seats in the parliament.

2. THE STIFFNESS OF THE TRADITIONAL PARTY SYSTEMS:
The Lipset-Rokkan Freezing of Party Systems Theory

The rise of Green parties in the European political scene has shattered one of the most well-known theories of the emergence and development of Western parties and party systems: The Freezing of Party Systems Theory.

The freezing of party systems theory was formulated by political sociologists Seymour Martin Lipset and Stein Rokkan in their remarkable 1967 essay "Cleavage Structures, Party Systems, and Voter Alignments." According to their essay, the seemingly great variety of twentieth-century democratic party systems can trace their development to two historical revolutions - the national and the industrial - and to two sets of social cleavages - the cultural, including differences in religion and ethnicity, and the economic, including differences between classes and between farmers and city dwellers. After the rise of the nation states and the industrial revolution, party systems "froze", to use Rokkan's and Lipset's metaphor. Therefore, contemporary democratic party systems tend to be organized around societal cleavages and political ideologies taken from the distant past.

Lipset and Rokkan demonstrated that the Western party systems of the 1960's reflected the cleavage structures of the 1910's and 1920's: *"An amazing number of parties which had established themselves by the end of World War I survived not only the onslaughts of Fascism and National Socialism but also another World War and a series of profound changes in the social and cultural structure of the polities they were part of.* [1]

2.1. The Frozen Swedish Five-Party Syndrome

Beside Michels' Iron Law of Oligarchy, the Freezing of Party Systems Theory is the most famous of all party theories. The Swedish constellation of five competing parties, frozen since the end of World War I and never thawed, seemed to be a perfect instance of this (Back & Berglund, 1978, Berglund & Lindström, 1978, Bäck & Möller, 1990).

The situation which prevailed on the eve of the 1988 election with five parties represented in Parliament - a Communist, a Social Democrat, a Liberal, an Agrarian, and a Conservative party - had been developed and clearly discernible since 1921.

Every representation in such a complex matter as a party system must be deficient, particularly if it purports to show changes. Figure 1 provides a sketchy overview of the emergence and development of the five-party system. The five traditional parties are indicated by solid lines.

2.3. The Five-Party System Froze Along the Left-Right Dimension

The inherited Swedish five-party system mainly froze along the well-known left-right dimension, which is construed as a model of ideological differences, of a cleavage in the world of ideas. Simplistically, left-wing ideologies bolster socialism, right-wing world views support capitalism. The former recommends the use of government to hamper what is considered economic exploitation of workers and clerks, to strengthen the position of consumers against producers and sellers, and to ameliorate the lot of tenants in relation to their landlords. The latter maintains that the government should be limited because public intervention impedes the natural play of market forces, reduces the freedom of individuals and inhibits their incentives to work, all of which damage the pursuit of the common well-being.

The location in the 1970's of the five major parties as organized, collective, national actors along the left-right dimension is illustrated in Figure 2. Closest to the left pole are the Left-Party Communists, followed by the Social Democrats and the Liberals (the People's Party), and closest to the right pole are the Conservatives (the Moderate Unity Party). These four political tendencies, and their organized expressions at the party level, exist in almost every European country west of the former so-called Iron Curtain.

2.3. The Five-Party System Also Froze Along the Center-Periphery Dimension

The existence of an Agrarian/Center Party points to another conflict axis around which the traditional Swedish five-party system has frozen, the urban-rural dimension or the centre-periphery dimension. Urban ideologies favour growth of the industrial sector in relation to the primary agricultural sector and support urbanization, whereas rural

---

[1]   Lipset & Rokkan 1967, 50 f. The Freezing of Party Systems Theory is my own term; the authors talk about the "freezing" of the systems, but do not use the full expression.

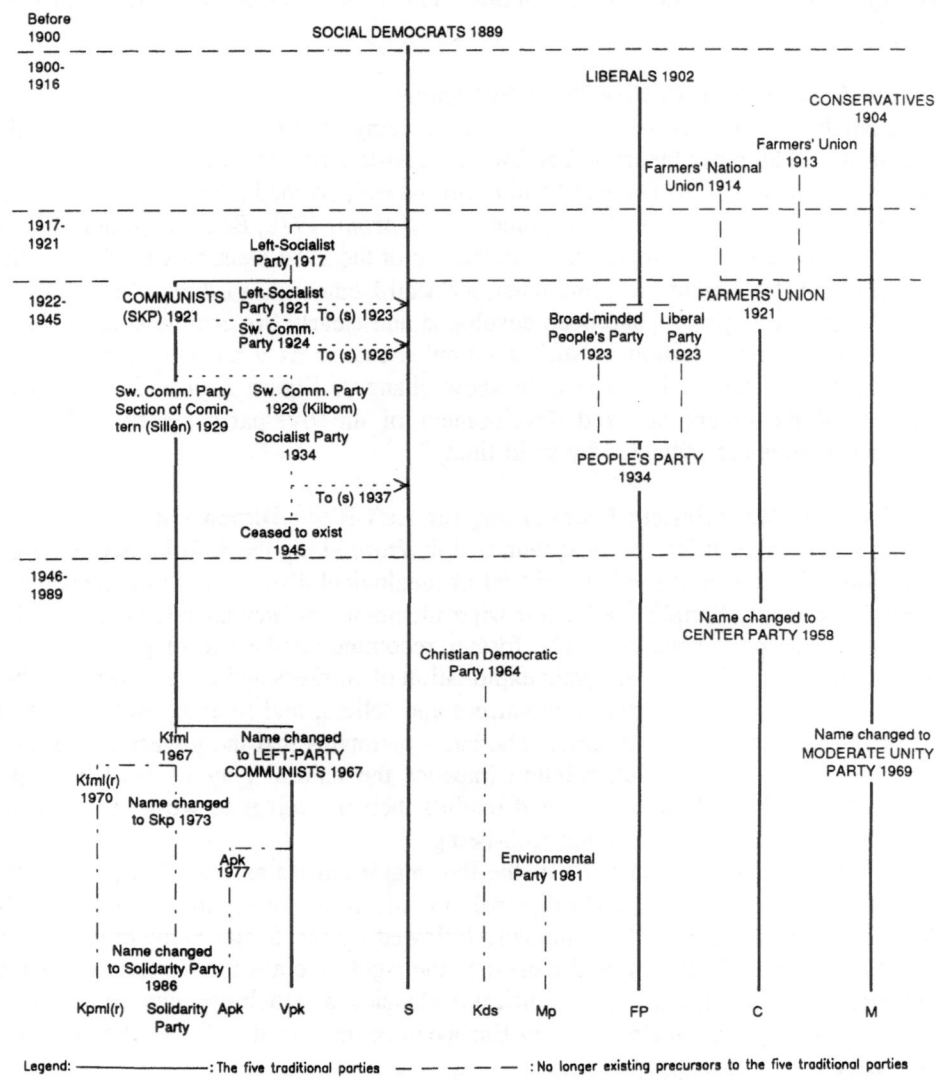

Figure 1. The freezing of the Swedish Five-Party System in 1921
and its development until 1988.

ideologies foster the interests of farmers and rural dwellers and champion "a living countryside". This dimension differentiates the Agrarian/Center Party from the other four.

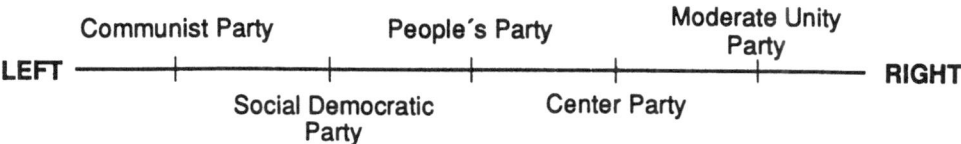

Figure 2. The Left-Right Political spectrum in Sweden in the 1970's.

3. WHY GREEN PARTIES?

The emergence of Green parties in most affluent, technological, service societies must be understood against the backdrop of a wave of ideological concerns that successively spread all over the Western world in the 1960's and 1970's. How this new wave of concerns should be characterized is hotly contested. Some observers regard it as environmentalism, a movement in favour of a cleaner world, with less pollution, less depletion of natural resources, conservation of wilderness areas, and recycling of resources. Other pundits prefer to conceptualize it as ecologism or post-modernism. They view the new ideology as a very broad reaction against the modernistic project encompassing industrialization, urbanization, and technological development. They also view it as in favour of ideas like limitations on growth, a sustainable society, and concern for the environment, not for human reasons but because the environment, with its animals and plants, has an intrinsic value. Still other scholars believe it involves post-materialism which, in addition to environmentalism or ecologism, includes the promotion of peace and disarmament, women's liberation, and increased popular participation in politics. Paehlke (1989, 3) has characterized environmentalism as "the first original ideological perspective to develop in the West since the middle of the Nineteenth Century". This is clearly an exaggeration. Environmentalism is the first original ideological perspective to develop in the West since Fascism in the 1920's.

Probably the most potent determinant promoting the rise of Green parties was the opposition to nuclear energy. Nuclear opposition was important in Germany, Australia, Austria, Sweden and many other countries.

Yet the upsurge of concern for the environment cannot alone explain the formation of Green parties. There are countries in the West with no Green party, Norway as the most notable example. The international wave of ecological ideas has not automatically produced Green parties in all Western countries. While necessary, environmentalism is not a sufficient factor supporting the ascent of Green parties.

Ideas count, but institutions decide. This allusion to Stein Rokkan's famous dictum *"votes count but resources decide"* captures our contention that ideas are important, but institutional response to the ideas is crucial for the formation of new parties. The way established institutions and, most notably, the traditional party systems react on new issues are crucial in the establishment of new political parties. If the existing institutional systems provide favourable opportunity structures for new popular movements, there will be no new parties.

As applied to Green parties, our theory can be phrased like this: If established party channels have not been able or willing to provide acceptable linkages between the new Green electorate and the leading echelons of governments, the chances for the formation of Green parties are greater (Kitschelt, 1986, Lawson, 1988).

The proliferation of Green parties is caused by the inability of the states, or more particularly the established parties and party systems, to accommodate the new set of environmental problems that slowly arose in the 1960's and 1970's. Of decisive importance was the response of the traditional parties to the rising popular anger against nuclear energy.

4. THE SWEDISH ENVIRONMENT PARTY:
Post-Materialism and Party System Failure

The Swedish Environment Party was founded on September 20, 1981.[2] Can the rise of this new party be explained by the Swedish version of the international wave of ecological, environmental or post-materialist ideas and considerations, in combination with the failure of the Swedish state, or more particularly the traditional five-party system, to channel these new ideas in the electorate into robust governmental decisions?

4.1. The State Responds Successfully to the Early Environmental Opinion
In the 1960's, the Swedish five-party system responded quite successfully to the rising environmental opinions in the populace.

In 1963, a State Nature Protection Council was created. Four years later, the new Environmental Protection Board brought together, in one body, the Nature Protection Council and all other state agencies that had previously dealt with environmental problems. *"It was the first of what would become new state environmental agencies in just about every country of the world,"* the Swedish sociologists Helena Flam and Andrew Jamison argued. *"Two years later, the Parliament passed an environmental protection law that was the strongest and most*

---

[2]   The formation of the party took place at a meeting in Örebro on September 18-20, 1981. - The initial Swedish name Miljöpartiet was changed in the fall of 1984 to Miljöpartiet De Gröna - literally: The Environment Party the Greens. See Weinberg, 1982, Gahrton, 1980, and Bennulf, 1990.

rigorous legislation that could be found anywhere. Many looked to Sweden and its pragmatic reformism as an exemplar in the new environmentalism" (Flam & Jamison, 1989, 13).

Largely because of its growing goodwill in the field, the Swedish Social Democratic government succeeded in persuading the U.N. authorities to pick Stockholm as the site of the world-wide Environmental Protection Conference in 1972.

The reason why the party system acted favourably upon the new environmental consciousness in the population was that it also had frozen around an old centre-periphery dimension. In other words, the explanation was to be found in the role of the Center Party in Swedish politics. Starting in the beginning of the 1960's, the Center leadership made great efforts to change its farmers and rural base into a more catch-all organization, fighting for decentralization and protection of the environment. The party leadership was actively searching for issues which could be used for this transformation purpose. This made it necessary for the others, the Social Democrats included, to show an interest in environmental issues in order to keep their votes and positions (Larsson, 1980).

Electricity policies posed the most serious problem to the five-party system.

4.2. The 1970 River-Saviors' Alliance: the First Disturbance in the Left-Right Pattern

The first electricity conflict was over the issue of harnessing the remaining wild rivers in the northern part of the country; of particularly great importance was the struggle concerning the Vindel River in the 1960's. Originally, the parties took their stances in conformity with the ingrained left-right pattern. The more the party leaned to the left, the more it favoured exploitation. The Communists and the Social Democrats were most in favour of harnessing, the Conservatives least.

However, the popular opposition became so strong that the Social Democratic government in 1970 had to abandon its plans to harness the Vindel River. It decided to exploit instead the so far untouched Kalix River. Surprisingly, the Social Democrats were grudgingly forced to retreat again, this time by an unexpected, unholy alliance of the Left-Party Communists and the three Nonsocialist parties.

The river-saviors' alliance in 1970 constituted a paradoxical novelty in Swedish party politics. Rarely, if ever, had Communists joined forces with Liberals and Conservatives against Social Democrats. For the first time, the new environmental concerns had demonstrated their capacity to create disturbances in the traditional Swedish Left-Right political spectrum. Far from being coincidental, the new river-saviors' alliance remained intact in connection with the 1972 and 1977 decisions concerning a system of national physical planning and the 1975 major energy policy decision (Vedung, 1984).

There were even clearer signs that the conventional parties had problems in mastering the situation. In 1968, popular action groups and movements suddenly mushroomed, particularly in the larger cities. They were local in their purview, acted on one singular issue at a time, and were critical of freeways, destruction of inner cities, and construction of new suburbs with high-rise housing. In due time, the single-issue action groups started to organize into umbrella organizations all over Western Europe. In Sweden, the Friends of the Earth and the National Association of Environmental Groups were both formed in 1971. In the same years, 1971 and 1972, similar organizations were formed in France, the Netherlands, and West Germany (Müller-Rommel, 1989, 6).

Another indication of the vulnerability of the conventional Swedish five-party system was that local Green or Alternative parties were formed, and gained some electoral support in several communities, throughout the country after 1972. The Alternative Party formed in 1979 in the country's capital and was named the Stockholm Party. It captured three seats in the municipal election which were held in the fall of that year and obtained substantial national media coverage.[3]

The growing peace and women's movements also kept clear of traditional party institutions.

The most serious threat to the traditional, frozen five-party system was another electricity issue, nuclear energy. The way popular anxiety towards nuclear power was handled by the inherited Swedish system of political institutions, particularly the Center Party in government between 1976 and 1978, and the Social Democrats, the Liberals, and the Moderates from 1973 until the 1980 referendum, was crucial in the formation of the Green Party in Sweden.

Nuclear energy has been of utmost importance to Alternative movements and Green parties in many countries. Even referenda on nuclear power have been crucial. *"What the Boston tea party was for the United States, the referendum of 1978 whereby the Austrian population decided to ban nuclear energy from Austria, was for the Greens in Austria,"* writes Christian Haerpfer on the development in his country (Haerpfer in Müller-Rommel, 1989, 23).

4.3. Nuclear Energy, the Growth-Ecology Dimension, and the Crisis of the Five-Party System

Sweden's nuclear energy program - allegedly the largest in the world on a per capita basis - unexpectedly became a subject of deep public controversy in the Spring of 1973, half a year before the Oil Crisis. The antinuclear sentiment did not grow upwards from the bottom but downwards from the top. The process through which the antinuclear issue was brought up on the political agenda did not start with local popular protests which slowly and grudgingly were accepted by some politicians or party at the center. On the contrary, it was rapidly adopted in the political center of the system, by some political parties, and from there it trickled down to the general public.

The reason why the antinuclear crusade started top-down and at such early moment in Sweden can be given an institutional explanation. From the end of the 1960's, the Center Party leaders attempted to transform their rural interest party into

---

[3]   The first local environment party was formed in Ängelholm in the southernmost province of Scania in 1972. Later, similar parties were established in Båstad, Sollentuna, Halmstad, Borgholm, Falkenberg, and Stockholm (February 4, 1979).

a party promoting decentralization and improved environment. The leadership actively searched for issues to use for this purpose. The 1973 adoption of the antinuclear cause is associated with this tendency.

Again, we see that the Swedish five-party system was exceptionally sensitive to environmental discontent because it contained a center-periphery dimension on which the old Farmers' Union, turned Center Party, occupied a position closest to the periphery pole.

The nuclear issue was difficult to handle for the other parties, however. The nuclear industry was very much the result of political planning. The Social Democrats in particular, but also the other parties, had promoted nuclear power, not only as a source of electricity, but to keep Sweden in the forefront among industrial nations. The nuclear power plants represented enormous investments. To dismantle them would mean waste of immense dimensions. Therefore, the Social Democrats, the Conservatives and the Liberals chose to proceed very carefully.

The bitter and prolonged controversy between advocates and enemies of nuclear energy created a growth-ecology cleavage alongside the traditional left-right dimension in the Swedish five-party system at the elite level. This ideological cleavage also structured the parties into a spectrum different from the conventional left-right pattern (see Figure 3).

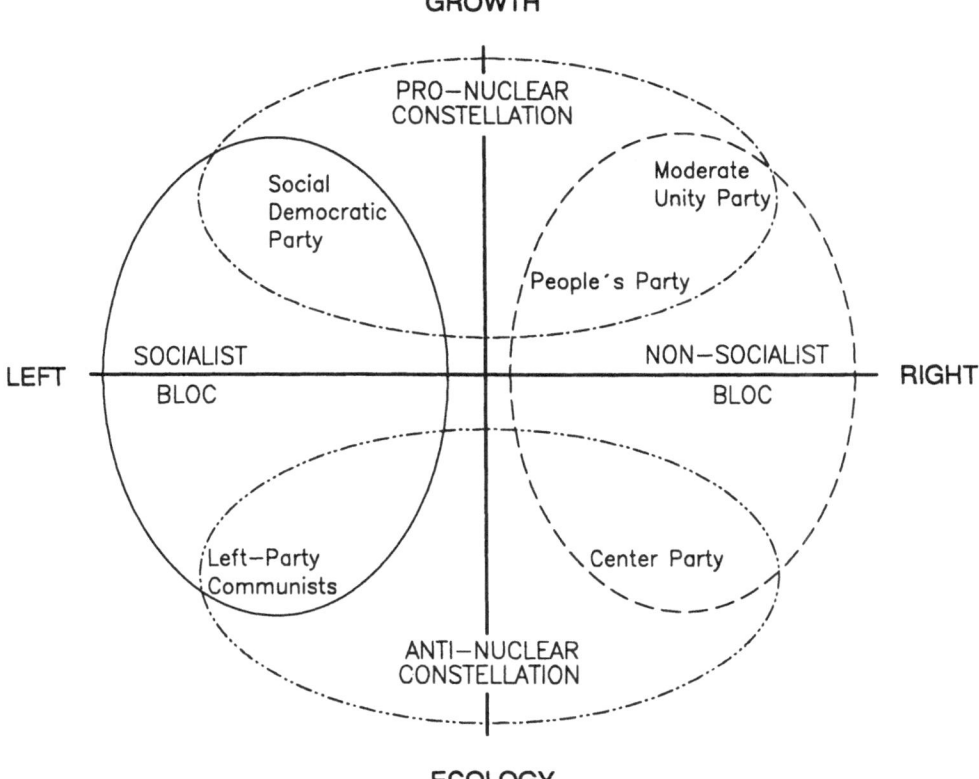

Figure 3: The Conflict Pattern in Sweden's Five-Party System in 1975-80.

Figure 3 reveals that the nuclear conflict, which occurred along the growth-ecology dimension, created tensions in the two traditional partisan blocs along the left-right axis. While the left-right cleavage indubitably remained predominant, the nuclear energy conflict produced uneasiness in the double-bloc constellation, pitting Communists against Social Democrats, and Centerists against Liberals and Moderates. Nuclear energy was the first great political issue to break down the habitual left-right party pattern in Swedish politics.

In addition, nuclear energy also created strong dissentions within the parties, in their electorates as well as among their ruling elites. All this showed the potentially explosive force of nuclear energy as a political issue.

The nuclear-antinuclear confrontation is conceived of here as part of a more comprehensive *growth-ecology dimension*. The reason is that nuclear power was discussed as the incarnation of the polluting, plundering, and materialistic industrial society. The debate did not center on nuclear power as a specific technique for the production of electricity but as the foremost symbol of the limits, risks and dangers of large-scale, state-supported technological development and continuing material growth. All kinds of ecological arguments were raised against nuclear energy.

The growth-ecology dimension concerns the rift between a pro-growth political ideology, arguing that the world is there to be used and exploited by man as a resource, and a world view averse to anything but the most timid engineering of the natural world by human beings. Another feature of the growth-ecology dimension concerns attitudes toward large-scale enterprises, centralist government planning and increasing bureaucratic control and coordination, where the ecology pole stands for the establishment of more decentralized communities in a repopulated countryside, in order to minimize environmental degradation and create a more self-sufficient and self-sustaining economy. This is captured in dictums like "small is beautiful", "steady-state economy", "sustainable society" and "zero-growth".

The growth-ecology dimension obviously differs from the traditional left-right dimension, as the political ideas involved do not concern the distribution of wealth or power among those who have and those who have not, within the general framework of continued industrialization. Instead, the growth-ecology dimension covers issues regarding the level and type of welfare, while debate along the left-right dimension concerns the distribution of welfare among different sectors and members of society. The growth-ecology dimension concerns an overriding feature common to both Communism and Capitalism, a super-ideology so to speak, namely industrialism.

4.4. The Center Party Shatters the Expectations for Nuclear Dismantling

Nuclear energy provoked an environmental political struggle with no counterpart in Swedish history. The doggedly antinuclear stance of the Center Party in the 1976 election campaign contributed to a political earthquake: the Nonsocialist forces captured a majority of the seats in Parliament and the Social Democrats were ousted from government. As a consequence, a three-party, center-right coalition cabinet was formed. For the first time in 44 years, the Social Democrats were in opposition. Nuclear energy had proved its disruptive force in traditional Swedish party politics.

Activists in the Alternative and Environmental movements hoped that the Center Party would phase out nuclear power. However, only supported by 24 per cent of the electorate, the Center Party leadership felt compelled to compromise with the other two, pronuclear Nonsocialist parties in order to seize upon the unique opportunity to place the Social Democrats firmly in opposition. In the following two years, the party leadership chose to compromise time and again to save the precious, three-party, Nonsocialist government. Saving the Nonsocialist government was more important, it seemed, than the dismantling of nuclear energy.

The Center Party's credibility as an environmental party was also undermined by the old nucleus of the party, the farmers'representatives. They were not particularly anxious to reduce the use of pesticides and herbicides in agriculture, in spite of strong environmentalist pressure. They also favoured a compromise on nuclear energy. It was not easy for the party leadership to throw away the party's rural past and open it up for modern environmentalists.

4.5. The Catalytic Factor: The Nuclear Referendum in 1980

Early in 1979, the political situation on nuclear energy changed radically in Sweden as a consequence of media response to a singular event in the international environment: the Three Mile Island nuclear incident in Pennsylvania in March.

Modern mass media are more than neutral channels for other people's opinions. They exert power by themselves. The media storm in the wake of Three Mile Island, more than the incident itself, compelled the pronuclear Social Democrats to agree to the long-standing demand by the Center and Communist parties to use an alternative decision mechanism, the referendum institution. The other parties followed suit. The national, consultative plebiscite was to take place half a year after the Parliamentary election in September. Again, nuclear energy had demonstrated its capability to pose problems to the established Swedish representative system of government.

The poll, which took place in March of 1980 when Sweden had six stations in use, four more ready to start up and two under construction, ended with a defeat for the antinuclear movement. The two pronuclear choices drew approximately 60 per cent, as compared to the nonnuclear's 39 per cent (Holmberg & Asp, 1984, Johansson, 1980). The result is displayed in Table 2.

Table 2
Results of the Swedish Nuclear Referendum, March 23, 1980.

Options	Supporting Parties	Number of Valid Ballots	Percentage of Valid Ballots	Turnout
No.1.	Moderate Party	904,968	18.9	
No.2.	Social Democrats People's Party	1,869,344	39.1	
No.3.	Center Party Communist Party	1,846,911	38.7	
Blank ballots		157,103	3.3	
Total		4,778,326	100.0	75.6

Option No. 1. Nuclear power is to be phased out... To reduce our dependency on oil,.. at most 12 reactors which are now in operation, completed or under construction are used. There must be no further expansion of nuclear power.

Option No. 2. Nuclear power is to be phased out... To reduce our dependency on oil,.. at most 12 reactors which are now in operation, completed or under construction are used. There must be no further expansion of nuclear power.
...The main responsibility for the production and distribution of electricity must be in public hands. Nuclear power stations and any other future installations of importance for the production of electricity must be owned by the State and the Municipalities.

Option No. 3. "No" to the continued expansion of nuclear power.
The phasing out of the six reactors now in operation within a period of, at most, ten years.

The debacle was hard to accept for the no-nukes activists. Even more discomforting was the sleazy behaviour of the established parties in connection with the referendum. The political ecologists in the Campaign were bitterly disappointed on three major accounts.
The machinations that took place before the plebiscite, when the five established parties had been negotiating the phrasing of the choices to be put forth, caused dismay and outright anger. The fact that two "pro-nuclear" alternatives were offered instead of one was considered an outrageous tactical manoeuvre by the Social

Democrats. As emerged from Table 2, the Social Democratic option (No. 2) was, for all practical purposes, identical to the one supported by the Moderates (No. 1), with the exception of an additional section regarding public ownership of the nuclear power plants. The antinuclear activists claimed that the latter, rather irrelevant demand had been consciously inserted to force the Moderates to frame an option of their own. They could not possibly support more Socialism. Through this manoeuvre, the antinuclear people felt they had to fight two pronuclear alternatives instead of one, which was considered more natural.

The part common to the two pronuclear alternatives was also a cause of dismay among antinuclear activists. Both options 1 and 2 included an initial statement that nuclear power would be phased out - this might be interpreted as a victory for the environmental opposition. However, before the close-down was to occur, the six additional reactors that were either finished or under construction would be taken into production to alleviate dependence on foreign oil. This would be in addition to the six reactors that were already in use. The so-called nuclear phase-out referred to in options 1 and 2 obviously implied more than doubling the actual Swedish nuclear capacity. This was debunked as dishonest double-talk designed to keep lukewarm antinuclear people from choosing the complete elimination, within ten years, of Sweden's nuclear capacity, as implied in the environmental opposition's option 3.

Finally, the antinuclear people were angered by the promises of all parties to abide by the result of the referendum. This was interpreted as a betrayal of the antinuclear cause for short-sighted tactical motives.

In the aftermath of the nuclear referendum, all five established parties stood out as fundamentally compromised in the eyes of the environmental activists. The established alternatives could not be counted on. Furthermore, the environmentalists had attempted to phase out nuclear power through the mechanism of ordinary parliamentary elections and through the extraordinary mechanism of referendum. Both had failed. The most promising remaining option was to form a party.

Yet, the plebiscite was also a catalyst for the party formation in a more constructive way. Swedish referenda are almost entirely controlled by the established parties, but through their extraordinary character, they provide opportunities that do not exist in ordinary election campaigns. The referendum forced the antinuclear activists to form a new organization, the People's Campaign Against Nuclear Power, in order to back up option three. Through experiences during the campaign, people realized that they could work together which paved the way for the party's formation in 1981.

Political entrepreneurship also played a role. Instrumental in the formation of the Environment Party was the former Liberal Member of Parliament, Per Gahrton, who after the referendum sent a personal manifesto to a number of people and some representatives of small parties having representatives on about 10 different municipal councils. A committee was established and on September 19-20, 1981, the Environment Party was formed as a national political party (Weinberg, 1982, Bennulf, 1990, Gahrton, 1980).

5. CONCLUDING REMARKS

Ideas count but institutions decide. In the present paper, I have emphasized the pivotal role played by the State and, particularly, by the traditional political institutions in the formation of new political parties. Stressing the importance of central political institutions, like political parties and party systems, does not engender the discounting of general, cross-national ideological trends and belief systems. It means that the importance of general ideological trends is translated through the national political party system. More specifically, according to the prime hypothesis advanced by Lawson (1988), alternative organizations are formed when major, established parties fail to provide acceptable forms of linkage.

Underlying the formation of the Green Party were three perceived electoral linkage failures, all of them related to nuclear energy.

Firstly, in office, the Center Party did not succeed in providing the ideological, antinuclear electoral linkage that the would-be founders of the Green Party desired. The Center Party failed to act as an intermediary capable of linking the antinuclear grass-roots opinion to national decision-making. The failure of an established and ostentatiously antinuclear party to effect any substantial change in Sweden's nuclear energy policy while in government was certainly one factor behind the emergence of the Green Party in Sweden.

Secondly, there was also a perceived deficiency in the referendum, an extraordinary tool provided by the inherited political system. In the eyes of the antinuclear activists, the plebiscite had failed to produce a satisfactory outcome due to manipulation by the pronuclear parties - the Social Democrats, the Moderates, and the Liberals.

Furthermore, the promise of all parties, and particularly of the Center Party and the Left-Party Communists, to abide by the results of the referendum, even though it would be pronuclear, was interpreted as a betrayal of the antinuclear cause.

The referendum was also important for another reason. The various organizations comprising the environmental movement had to join forces in the Campaign against Nuclear Energy. In working together, they discovered their similarities and created a social base for a party.

ACKNOWLEDGEMENTS

The present article, finished in October of 1990, is part of a continuing effort of mine (Vedung, 1988, 1989) to shed some light on Green ideas and Green parties. My warm thanks to Erik Ib Schmidt, member of the Symposium Programme Committee, for his valuable comments, to system analyst Peter Knutar of Uppsala Data Center, for drawing some of the figures, and to Laila Grandin in the Uppsala Department of Government for daily technical assistance.

REFERENCES

Back, P. & S. Berglund (1978), Det svenska partiväsendet, Stockholm: Almqvist & Wiksell.

Bennulf, M. (1990), Grönt ljus: Väljarna och miljöpartiets valframgång 1988, Gothenburg University, Dept of Political Science: mimeo.

Berglund, S. & U. Lindström (1978), The Scandinavian Party Systems: A Comparative Study, Lund: Studentlitteratur.

Bäck, M. & T. Möller (1990), Partier och organisationer, Stockholm: Allm. Förlaget.

Flam, H. & A. Jamison (1989), "The Swedish Confrontation over Nuclear Energy: A Case of a Timid Antinuclear Opposition," to be included in the Scass study of antinuclear movements and the state. Uppsala: Swedish Collegium of Advanced Study.

Gahrton, P. (1980), Det behövs ett framtidsparti, Stockholm: Prisma.

Holmberg, S. & A. Kent (1984), Kampen om kärnkraften: En bok om väljare, massmedia och folkomröstningen (1980, Stockholm: Liber.

Johansson, L. (1980), Kärnkraftsomröstningen i kommunerna, Lund: Student litteratur.

Kitschelt, H. (1986), "Political Opportunity Structures and Political Protest: Antinuclear Movements in Four Democracies," British Journal of Political Science, 16: 57-85.

Larsson, H.-Albin (1980), Partireformationen: från bondeförbund till centerparti, Lund: Liber.

Lawson, K. (1988), "When Linkage Fails" in K. Lawson & P. H. Merkl, eds., When Parties Fail: Emerging Alternative Organizations, pp. 13-38, Princeton, NJ.: Princeton University Press.

Lipset, S., M. & S. Rokkan (1967), Party Systems and Voter Alignments: Cross-National Perspectives, New York: Free Press.

Müller-Rommel, F., ed. (1989), New Politics in Western Europe: The Rise and Success of Green Parties and Alternative Lists, Boulder, CO: Westview Press.

Parkin, S. (1989), Green Parties: An International Guide, London: Heretic Books.

Vedung, E. (1984), "Striden om de strömmande vattnen," Tekniska Museets Årsbok Daedalus, pp. 105-161, Stockholm: Tekniska Museet.

274   E. VEDUNG

Vedung, E. (1988), "The Swedish Five-Party Syndrome and the Environmentalists" in K. Lawson & P. H. Merkl, eds., When Parties Fail: Emerging Alternative Organizations, pp. 76-109, Princeton, NJ.: Princeton University Press.

Vedung, E. (1989), "Sweden: The 'Miljöpartiet de Gröna'" in F. Müller-Rommel, pp. 139-153.

Weinberg, G. (1982), "Hur miljöpartiet växte fram", in Nu kommer Miljöpartiet: Om Miljöpartiet av miljöpartister med det officiella partiprogrammet, pp. 17-27, Stockholm: Timo Förlag.

17.

URBAN GREEN OASES AND RECREATIONAL AREAS

DIETER MAGNUS
Environmental Artist, Bingerstrasse 4,
6501 Wackernheim, Germany

ABSTRACT

The most important task for redevelopment in the next decades everywhere in the world will be the improvement of living conditions in big cities. The revitalizing of old and new city districts and residential quarters, as well as the redesigning of squares, courtyards and gaps between buildings have all taken on a new significance.

Urban repair is a complex task requiring fantasy, an intuitive grasp and modesty; it should not be restricted solely to the traditional professions.

Urban renewal, even from the artistic point of view, is still largely misunderstood. Environmental art is something different, something going farther - related to a certain town planning situation and to the people living there. It is not isolated works of art but forms of integration, always related to one another. Environmental art is also an openended process allowing continuity and change. With the reconstruction of squares, courtyards and human-dimension intermediate spaces, the prerequisites can be created for a more selfdetermined local and neighbourhood culture. The artistic design of such public, and also private, areas is of great importance because it can create surroundings which stimulate one's fantasy and make participation possible.

Key Words: Environmental Architecture, Environmental Art, Green Oases, Overall Concepts, Recreational Areas, Urban Art, Urban Ecology, Urban Space.

ENVIRONMENTAL DESIGN
- Reference to Nature and Environmental Art

"The city should be a place where socialization and communication are possible. It should provide incentive for individuals to act and think in a creative manner. This implies that municipal development should not be based entirely on sober functional consideration but should include the element of artistic intention". (Quotation from: Empfehlung des Deutschen Städtetages zu "Kunst und Bauen", Berlin, 1974). This quotation from the German Cities' Assembly goes way beyond the usual inclusion of art in structural design.

Meanwhile, more than 15 years have passed by, this statement still remains insufficient, and the unsatisfactory role of art in urban areas has actually not progressed. The artist working in the public sector should consider social and ecological aspects in addition to aesthetic view points. Facing an environmental and urban crisis, it is necessary to be far-sighted and to work progressively, thus, making a new line of orientation indispensable. The aim is to learn from the mistakes of the past, to formulate new priorities, to win back a human scale, to oppose the further destruction of Nature and to emphasize the regional cultural inheritance which creates a sense of identity into urban design. There is an urgent need for urban repair almost everywhere and this need must be met with an approach towards nature and human design. It should gradually reawaken an appreciation of and a sharpening of the senses, introducing new aesthetics into our way of life and our inter-social relationship.

Therefore my keynotes: Reference to nature or experience of nature as an element of renewal and survival, environmental art as a form of integration and process are important suppositions for regaining and using urban spaces.

I am neither an architect nor a town planner, but rather a painter and sculptor who, through bewilderment and discontent with badly-built surroundings, gradually became an environmental artist searching for a path towards joint design and responsibility in the urban area.

The possibilities for the artist between the traditional professions of architect, engineer, designer and landscape gardener are greater than first appearances would suggest. Independent of the official art scene, there are developments at an international level which are changing the traditional role of the artist in the urban sphere: artists with new activities and abilities are becoming partners of town planners and citizens.

The design of city areas from an artist's point of view must be preceded by new thinking by planners, politicians and not least of all by the artists themselves. Today the role of the artist in the urban sphere must also be viewed from social aspects - not for the artist's sake, but rather for his possible achievements to society in the sense of a newly formulated quality in the environment.

Not only is the structural environment, which will be present in the year 2000, already existing in three quarters of Europe; it will make it unlikely that new ideas will get a chance to overcome conventional thinking. Therefore, it is imperative to destroy the blinker mentality of the planning disciplines so that the problem can be approached through cooperation. It must be our goal to begin a process where we, by means of new ideas, will have a chance to achieve a new and better environment.

An important task for redevelopment in the next decades everywhere in the world will be the improvement of the living conditions in big cities. This statement is valid not only for the industrialized countries but to an even larger extent for the developing countries. Unrestrained forms of settlements, the development of car traffic and the world-wide extension of the so-called "international modern style" in architecture, often have destroyed grown structures and cultural identities. The restoration of old and modern dwellings as well as areas in the city, and the redesign of detailed areas such as squares, yards, and empty lots have a new priority. The importance of the quality of the design in small open areas has been underestimated or even ignored. It is the artist's chance as well as responsibility in discussions with those concerned, to fill in these neglected areas with new ideas.

The artistic and natural design of such public places is of great importance since it can create surroundings which stimulate fantasy and make participation possible; by this I do not mean museum or studio art nor the usual sculpture on a pedestal but, rather, overlapping concepts for a complete nature in the urban residential environment, including street furniture and green zones. It is essential to search for alternative solutions in order to arrest the continuing flight from the inner cities and the progressive destruction of nature, which is taking place.

A poverty of forms in the city environment on the one hand and noise pollution on the other hand have negatively influenced the behaviour and development of the human being. The squalid state of public areas in the inner cities is an indication of the existing lack of both animation and possibility for activity.

The survival of big cities will be determined in part by how far we succeed in broadening the space in which people can act freely, participate and grow personally. As a result of the rebuilding of public areas more self-determined suburb cultures, more neighbourhoods, and a more active occupation, can be achieved. Therefore, the state and local governments must stimulate the development of participation and social behaviour through these pilot projects.

A memory of traditional values is unfortunately often pursued, not with regard to content, but rather as a matter of form with fashionable "-isms"; a lack of new ideas and modern messages remains. Architectural history, however, has many examples which demonstrate the residential area which can utilize the close interaction between art and nature.

Involving the residents concerned in the planned redesign of a residential area, and to combine this, where possible, with a revival of the suburbs culture, is an important prerequisite for the success of such projects. Unfortunately this vital prerequisite is often neglected by planners and those who commissioned the project.

Follow-up programmes are also essential to encourage the necessary adjustment, interest and acceptance of the residents.

The simple fact that the human being is part of nature is often ignored. This might have occurred because, especially in urban areas, many people do not realise their reliance on, and their bonds with, nature any more.

A quotation from "The Eight Deadly Sins of Civilized Humanity": *"The beauty of nature and the beauty of the cultural surroundings created by man are ... necessary in order to keep human beings intellectually and mentally healthy. Total psychic blindness to all beauty ... is a mental disorder which must be taken seriously, not least because it is accompanied by an insensitivity towards the ethically reprehensible. For those who have to decide whether a road, a power station or a factory is to be built ... aesthetic consideration do not play any role at all. There is complete unanimity from the council chairman of a small village right up to the minister of economics of a major state that no economic - or even less political - sacrifices should be made for natural beauty. The few conservationists and scientists who have their eyes open for the impending disaster are completely powerless".* (Konrad Lorenz, 1973)

Overcoming this blockage, put up by visual stencils and behaviour patterns, it is important to recapture the ability to experience nature and thus motivate creativity. Spontaneous discovery happens primarily in those areas which have not yet been occupied with a fixed meaning. For sometimes it is not the smooth facades which give an impetus for imagination but the gutter after a fall of rain or the cracks in an old wall. Occurrences which take place outside our sterile, ordered and organised city environment because what is apparently redundant, playful and decorative is necessary as an enrichment and impetus to human nature.

To preserve and regain the ecological essentials of life are the most important tasks for our generation. It is not enough to demand that others save the rain forests while you yourself sit at home putting your head in the sand in order to avoid the pressing problems of road traffic or the increasing overdevelopment of life which is endangering of soil, air and water because of pollution. It is not only politicians and city planners but all of us which have suppressed this time bomb. Radical change, rethinking and alternatives, therefore, are urgently needed in all spheres, not least in the urban sphere.

RECONSTRUCTION AND REUSE

As an artist I have been occupied for years by design of the immediate residential area and have had the opportunity of testing such environmental design in practice which has developed my ideas further. The relationship between constructed form and vegetation, and between art and nature, not as a formal-aesthetic principle but as a continuous process of development in the overall concept as well as in detail is of central importance.

My projects in Mainz, Melbourne and Glasgow in the following presentation will attempt to demonstrate this interplay between urban green-space and environmental art. The projects in Mainz, "Green Bridge" and "From Spacing Verdure to Residential Garden", and the project in Glasgow, "People's Square", also offer an example of how to sensitise and involve the inhabitants. The usual pattern which one is confronted by is: initial opposition from those immediately concerned and from would-be experts at the onset. Then comes the process of forming opinions through discussions at numerous public meetings and then after that the immediate personal on-site experience which lead to the gradual acceptance as an open-ended process for me. These are the essential prerequisites for the renewal and reuse of urban living spaces; urban culture and the artist are closely concerned with this.

1. Green Bridge - Artistic and Natural Landscape in Mainz, 1981: Reuse by Bridging a Public Highway.

If one has had the good fortune to be able to develop a structure, such as the "Green Bridge", over a long period of time from the large model to the waterfall, lamp, paving or planting, and if people accept this newly gained space bit by bit, then one gains an understanding of what the unity of the arts and life in public open spaces once used to mean.

Many of the bridges in cities are purely practical structures and more repellent than inviting. The "Green Bridge" was and is meant to be more. A section of the City of Mainz, called New City with 30.000 inhabitants, was divided from a recreational area near the Rhine River by a highway (Plate 1). To me, the job consisted of closing the gap between the two areas and, at the same time, offer the residents an attractive space. In 1977, I began with an idea which received a lot of discussion. After a good deal of political compromises and after a thorough planning, the construction work began in January of 1980 and was completed in May of 1981. The "Green Bridge" consists of a broad, multi-level platform with green hill landscapes and colour, corner seats and comfortable ramps. It also has play niches in the form of a waterfall and an open-air stage - a multi-functioning bridge structure (Plates 2-4).

Plate 1. Green Bridge-Project. A photograph of the former urban
situation. Mainz, Germany. (D. Magnus).

Plate 2. Green Bridge-Project. Present urban situation. Mainz, Germany. (D. Magnus).

Plate 3. Green Bridge-Project. A green hill in the middle of the town connecting a section of the city with a recreational area near the river Rhine in Feldberg Strasse. Mainz, Germany. (D. Magnus).

Plate 4. Green Bridge-Project. The Green Bridge waterfall is a great attraction in the summer to both young and old. Mainz, Germany. (D. Magnus).

"No way leads past the recognition that the form our cities have taken on in the industrial age no longer meets their inhabitants' basic vital and cultural require-ments. The object is to work for alternative solutions in order to counteract the exodus from the cities and to make inner-city living more attractive. The "Green Bridge" is an important step in this direction. It is piece of a new urban landscape, inviting and encouraging use". (Jockel Fuchs, former Mayor of the City of Mainz).

2. Batman Park and Urban Culture - a proposal of urban renewal in Melbourne, Australia, 1986-88, in collaboration with the Victoria Ministry of Planning and Environment, the City of Melbourne and the Goethe Institute: Reuse of a River Scape as a Recreational Area.
For the urban development and renewal of Melbourne, the areas along the banks of the Yarra are of great importance for local recreation. As a first step, the reconstruc-tion of Batman Park was planned. The latter is intersected by road and rail viaducts. The present state: the river site with ugly overfly and heliport (Plate 5).

The possible state with the design objectives: a continuous green landscape which, by means of intensive planting, terrain modelling and bridging over, will push back the ugly dominance of the traffic routes and offer the city dwellers attractive and sheltered areas to spend their leisure time - an avenue for pedestrians and cyclists, an exhibition pavilion and cafe, a waterfall and island, small hills, and seating niches;

Plate 5. Batman Park, Melbourne, Australia, in its present state: The river side
with its ugly overfly and heliport. (D. Magnus).

Plate 6. Batman Park, Melbourne, Australia, a possible future state: A continuous
green landscape which offers the city dwellers attractive and sheltered
recreational areas. (D. Magnus).

a continuous green scape which encompasses all the prerequisites for a varied urban culture: recreation, art market, events, shopping, barbecuing, etc. (Plate 6, opposite page). After several discussions among politicians and planners took place, a majority decided the area should be used for more building construction and commercial purposes. The idea of a green oasis combined with cultural events in the heart of Melbourne was thus rejected.

3. From Spacing Verdure to Residential Garden - Lodging Area, Mainz, Germany, 1989: Reuse by Redesign.

The old cities, which we so admire nowadays, had to achieve high housing densities in the smallest space within their fortifications, coupled with intensive planting which played an important role in the residents' modest self-supply of food. The greatest humiliation a foreign conqueror could inflict on a city in the Middle Ages was to cut down its fruit trees because this wound took decades to heal. The emphasis on such private open spaces, coupled with an expansion of social contacts, formed the basis of the garden city ideas of the turn of the century. In retrospect, there is no sensible reason why this humane building tradition involving residential courtyards and residential gardens should have been replaced after the Second World War, especially in Europe, by free-standing blocks and towers with sterile spaces in between.

The residential estate at Südring in Mainz was suitable as a model project in several respects. The blocks did not have any special significance and required a thorough rehabilitation. The generously dimensioned open space is boring and unused - and games are forbidden (Plate 7).

Plate 7. Spacing Verdure - Residential Estate at Südring, Mainz, Germany, in its former state: The space is boring and unused. (D. Magnus).

On behalf of a council flat company (Wohnbau Mainz) I did two things: Redeveloped the five houses with hipped roofs, new windows and a new design of the facades. I also redesigned the surrounding area to include a garden pond, playgrounds for the children and allotments for the tenants (Plate 8).

The objective was to give the area a new identity, to improve the residential quality, to safeguard the tenants in the long term, to make more neighbourliness possible and to offer recreational space for the tenants's own amusement. The redesign was carried out in close collaboration with the residents.

Plate 8. From Spacing Verdure to Residential Garden, Südring, Mainz, in its present state: A garden pond, play grounds for the children and allotments for the tenants. (Dieter Magnus).

4. People's Square - Glasgow 1990, Cultural Capital of Europe: Reuse of a Gap by Renovation and New Design.

April, 1989, saw the start of this project and the beginning of a unique cooperation between the Goethe Institute, several offices of the City Council and local community groups. A largely derelict gap site owned by the City of Glasgow was chosen to be transformed into a playground and recreational area. The site was partially neglected and is situated in Garnethill, an inner city residential area with a lively multi-cultural community, in the immediate vicinity of the Glasgow School of Art and close to other important cultural institutions (Plate 9).

Plate 9. People's Square, Glasgow, 1990. Former state of the site: Ugly and partially neglected. (D. Magnus).

The reshaping will offer the people who live there an exciting as well as an ordinary living space which can be used as a meeting point and as a landscape in which to play and relax. It is the aim that it shall showcase the city's cultural life (Plate 10). The 200 meter long creek with its branches gives the slightly sloping area its character by joining the different parts of the square: a meadow for games, a stone pyramid linked to a stone landscape and a changeable waterplay. Its tree houses will be made from dead elm trees, an amphitheatre with sandstone seating from demolished buildings and an open air stage with a glass pavilion will also be created. The work is now in progress, (Plate 11), and the opening festival will probably take place in July 1991 with events and theatre which the residents themselves will organize.

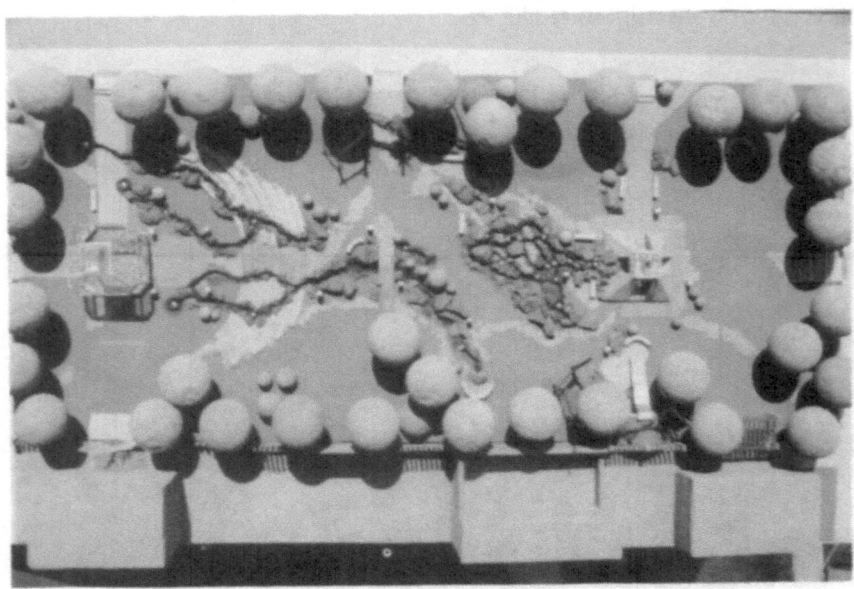

Plate 10. People's Square, Glasgow, Scotland. Model showing the design elements: Meadow, pyramid, stone and water landscape, amphitheatre and pavilion. (D. Magnus).

Plate 11. People's Square, Glasgow, Scotland. The work of reshaping the area is progressing, January, 1991. The form of the amphitheatre, creek beds and stone pyramid can be detected (D. Magnus).

CONCLUSION

And now for my conclusion. I have tried to make the close relationship clear which nature, environmental art and the renewal and reuse of urban spaces have with one another. I believe that urban green-spaces and urban culture are interdependent.

This is not a question of major projects but of the sum of many small steps in the interrelation of shaping, social behaviour and experience of nature - a process which has to begin everywhere and which has to develop further. The regaining of this sensitivity towards nature promotes the rethinking and the awareness of ecological necessities.

And the artist, as an expert in sensory perception and visualization, can make important contributions by providing fresh ideas, new techniques and skills. I can only concur with what the philosopher and physicist Klaus Michael Meyer-Abich wrote in his book "Ways to Peace with Nature": *"I consider the garden to be the cell for the renewal of our culture. The renewed perception of nature in the garden can take on form in a new aesthetics in everyday life, and point the way for a return to life for the industrial system"* (Meyer-Abich, 1984).

The survival of the human race on this planet needs a solidarity which will overcome frontiers and old ways of thinking. It is needed not tomorrow but now, because the war which people have declared on nature everywhere has already struck back.

REFERENCES

Empfehlung des Deutschen Städtetages zu "Kunst und Bauen", Berlin, (1974), Informationsblatt, p. 2.

Lorenz, K. (1973), "Die Acht Todsünden der zivilizierten Menschheit", R. Piper & Co. Verlag, Munich, p. 31.

Meyer-Abich, K.M. (1984), "Wege zum Frieden mit der Natur", Carl Hanser Verlag, Munich, Vienna, p. 267.

INDEX OF CONTRIBUTORS

SUBJECT INDEX